Secular Discourse on Sin in the Anthropocene

Environment and Society

Series Editor: Douglas Vakoch

As scholars examine the environmental challenges facing humanity, they increasingly recognize that solutions require a focus on the human causes and consequences of these threats, and not merely a focus on the scientific and technical issues. To meet this need, the Environment and Society series explores a broad range of topics in environmental studies from the perspectives of the social sciences and humanities. Books in this series help the reader understand contemporary environmental concerns, while offering concrete steps to address these problems.

Books in this series include both monographs and edited volumes that are grounded in the realities of ecological issues identified by the natural sciences. Our authors and contributors come from disciplines including but not limited to anthropology, architecture, area studies, communication studies, economics, ethics, gender studies, geography, history, law, pedagogy, philosophy, political science, psychology, religious studies, sociology, and theology. To foster a constructive dialogue between these researchers and environmental scientists, the Environment and Society series publishes work that is relevant to those engaged in environmental studies, while also being of interest to scholars from the author's primary discipline.

Titles in the Series

Secular Discourse on Sin in the Anthropocene: What's Wrong with the World? by Ernst M. Conradie

Environment, Social Justice, and the Media in the Age of Anthropocene, edited by Elizabeth G. Dobbins, Maria Lucia Piga, and Luigi Manca

Global Perspectives on Eco-Aesthetics and Eco-Ethics: A Green Critique, edited by Krishanu Maiti & Soumyadeep Chakraborty

Secular Discourse on Sin in the Anthropocene

What's Wrong with the World?

Ernst M. Conradie

LEXINGTON BOOKS
Lanham • Boulder • New York • London

Published by Lexington Books
An imprint of The Rowman & Littlefield Publishing Group, Inc.
4501 Forbes Boulevard, Suite 200, Lanham, Maryland 20706
www.rowman.com

6 Tinworth Street, London SE11 5AL, United Kingdom

British Library Cataloguing in Publication Information Available

Library of Congress Cataloging-in-Publication Data

Names: Conradie, E. M., author.
Title: Secular discourse on sin in the anthropocene : what's wrong with the world? / Ernst M. Conradie.
Description: Lanham : Lexington Books, 2020. | Series: Environment and society | Includes bibliographical references and index.
Identifiers: LCCN 2020033321 (print) | LCCN 2020033322 (ebook) | ISBN 9781793635075 (cloth) | ISBN 9781793635082 (epub)
Subjects: LCSH: Ecotheology. | Sin—Christianity. | Human ecology—Religious aspects—Christianity. | Environmental degradation—Religious aspects—Christianity. | Apartheid—Religious aspects—Christianity.
Classification: LCC BT695.5 .C6744 2020 (print) | LCC BT695.5 (ebook) | DDC 261.8/8—dc23
LC record available at https://lccn.loc.gov/2020033321
LC ebook record available at https://lccn.loc.gov/2020033322

™ The paper used in this publication meets the minimum requirements of American National Standard for Information Sciences—Permanence of Paper for Printed Library Materials, ANSI/NISO Z39.48-1992.

Contents

Acknowledgments

I would like to express my gratitude to the following persons and institutions that helped in the completion of the manuscript:

- To my colleagues in the Department of Religion and Theology at the University of the Western Cape for their support and friendship.
- To the Institute for Environmental Studies at the Vrije Universiteit Amsterdam, my colleagues Philipp Pattberg and Gijsbert van den Brink for offering me a fellowship on "The Ethics of the Anthropocene," and to Jan Jorrit Hasselaar and the interdisciplinary group of scholars who met to discuss early drafts of the manuscript.
- To participants in the colloquiums on "Redeeming sin?" hosted in Denver (2018) and in San Diego (2019) for critical feedback on aspects of the argument.
- To Hilda Koster and her colleagues at Concordia College (Moorhead, Minnesota) for inviting me for the Oen Lecture in November 2018 that helped me to clarify my thinking on "Whose Anthropocene – Which Diagnosis."
- To the Ethics 735 class of 2019 at UWC that joined me on the journey of reflecting on the Anthropocene from the perspective of the global South.
- To the team at Lexington Books for their efficient and professional service.

In addition to various references, in particular sections of this manuscript I revised portions of various previous publications for which I hold copyright and made use of the work submitted for publication elsewhere while working on the manuscript. Such cases are indicated through references in the text below where appropriate. The following instances may be specifically mentioned in this regard:

- "Is it not God's Mercy That Nourishes and Sustains Us . . . Forever? Some Theological Perspectives on Entangled Sustainabilities." *Scriptura* 116, 38–54. (See the section on "Shifts in Discourse on Sustainability.")
- "Why, exactly, is climate change a wicked problem?" *Philosophia Reformata* (forthcoming, 2020). (An extended version of the section on "A wicked problem?," written simultaneously.)
- "Is Christian Humanism what is needed in the Age of the Anthropocene?" *Stellenbosch Theological Journal* 5:3, 41–58 (a letter addressed to John de Gruchy for a special edition in his honor, for which I made use of some material in the section on "Humans in the Center? Four contrasting positions," written simultaneously).
- "Apartheid as Quasi-soteriology: The Remaining Lure and Threat." *Journal of Theology for Southern Africa* 138 (2010): 112–23 (coauthored with Murray Coetzee) (a few paragraphs in the section on "Apartheid as Quasi-soteriology").
- *Christianity and a Critique of Consumerism: A Survey of Six Points of Entry* (Wellington: Bible Media, 2019). (See the sections on "Idolatry and the Critique of Consumerism" and on heresy in the Anthropocene.)
- "Globalisation, Consumerism and the Call for a *Status Confessionis*," in *Globalisation Volume II: Global Crisis, Global Challenge, Global Faith – An Ongoing Response to the Accra Confession*, edited by Allan A. Boesak and Len Hansen (Stellenbosch: SUN Press, 2010), 53–76. (See the section on "Apartheid as a heresy.")
- "The Story of God's Work: An Open-ended Narrative." In Ernst M. Conradie and Hilda P. Koster (eds), *T&T Clark Handbook on Christian Theology and Climate Change* (London et al.: T&T Clark, 2019), 462–66 (an abbreviated version of parts of the chapter on "Apocalyptic Theological Postscript").

Dedicated to Jan Botha, for thirty years of sustained friendship.

Introduction

This contribution explores the question as to what Christian theology can contribute toward an adequate, multidisciplinary diagnosis of what has gone wrong in the world in the context of the Anthropocene. It assumes a notion of sin-talk as a form of social diagnostics as proposed in my book *Redeeming Sin? Social Diagnostics amid Ecological Destruction* (Lexington 2017). Here I will suggest a conceptual toolkit derived from the theological critique against apartheid in South Africa in the 1970s and 1980s in order to offer an authentic contribution to global discourse on the Anthropocene from within the global South. This toolkit distinguishes between different layers of what is wrong with apartheid, namely, as unjust cultural habits, as structural violence, as ideology, as a quasi-soteriology, as idolatry, and as heresy. By reviewing *secular* literature on the Anthropocene, I will seek to establish whether and to what extent such a theological toolkit can help to diagnose what has ultimately gone wrong with the world.

There are two personal and institutional contexts that serve as a background within which this contribution emerged. The one is a collaborative research project entitled "Redeeming Sin? Hamartiology, Ecology and Social Diagnostics" registered at the University of the Western Cape where I am based, while the other is a senior fellowship on "The Ethics of the Anthropocene" that I was privileged to hold at the Vrije Universiteit Amsterdam in 2018. Both are important in framing the argument of this contribution.

REDEEMING SIN?

The project on "Redeeming Sin?" emerged from the unbundling of the larger Christian faith and the Earth project[1] that identified the need for further reflection on the reinterpretation of the Christian notion of sin in the context of ecological destruction. This led to a series of intriguing international colloquiums on "Redeeming Sin?" between 2015 and 2019, a few postgraduate projects and has also been the main focus of my own research engagements since 2014.

The intuition behind the project is that Christian discourse on sin may be viewed, at least from the outside, as a form of social diagnostics. It offers a contribution to the colloquial but also deeply disillusioned question: What has gone wrong with the world?[2] It is a question raised by The Black Eyed Peas in rap but also by many others. Amid ecological destruction and the recognition of the advent of the Anthropocene, this question carries ominous, even apocalyptic overtones.

To speak of sin as a form of social diagnostics requires retrieval of sin-talk in the public sphere. This is in line with the prophetic tradition of social critique for the sake of justice but has become counterintuitive given the privatization and self-marginalization of sin-talk. All too often sin is reduced to matters of masturbation, sex before marriage, smoking and drinking (sin tax), a yellow card in sport leading to time in the "sin bin," eating "sinfully delicious" foods, sugar tax, or carbon emissions that may be "redeemed" through buying "indulgences" in some or other carbon trade scheme.

In response, viewing sin as a form of social diagnostics recognizes the need for multidisciplinary collaboration in order to fathom what is wrong with the world.[3] The terminology is derived from the disciplines of medicine and psychology, but this is also a task to which social theorists, economists, political analysts, environmentalists, philosophers, ethicists, artists (the "antennae of society") and, I argue, theologians can contribute. That there may be a role for theology here is born from prophetic discernment and from pastoral sensitivity but may also be understood as a form of Christian responsibility. Arguably, while some disciplines describe the symptoms, others seek to understand the underlying root causes of what is wrong. Such root causes are further radicalized (radix = root) in theological discourse to fathom what is *ultimately* wrong with the world. Although the terminology is derived from the fields of medicine and psychology where patients are not to be blamed for their illness, the notion of social diagnostics suggests that moral connotations are not avoided so that moral agents are not merely regarded as "patients" in need of medication and therapy.

As I will argue below, there are some conceptual tools that emerged through centuries of theological reflection that may be helpful in the public sphere, for example, to explore the root causes of ecological destruction, not least in the African context.[4] I have also employed such tools to engage with the National Development Plan (NDP) in the South African context.[5]

Given the cultural ridicule expressed toward sin-talk, any retrieval of discourse on sin in the public sphere faces some major obstacles. These obstacles shaped my own research agenda over the past few years, as the endnotes below indicate. Two examples indicate the complexity of the task:

First, there is a need to account for the emergence of sin (the so-called fall of humanity) in evolutionary history.[6] The "origin" of sin is often confused with the notion of "original sin" (which in my view is better understood in terms of the widespread, inescapable consequences of sin), but there are some thorny questions that have to be addressed on the relationship between social evil and natural evil, not least regarding the role of natural disselection,[7] aggression, predation,[8] and procreation. In my view, this need not be an insurmountable problem for the kind of a collaborative social diagnostics envisaged here. The Christian confession of sin expresses a conviction that is widely held, namely, that things are not what they are supposed to be,[9] what they could have been and should have been. Something went awry somewhere somehow. And when the world is not as it ought to be, we have every reason to ask why.

This would be contested only by those who accept some form of determinism (the mess in which we find ourselves is either necessary or inevitable), by those who are quite satisfied with the status quo, despite whatever injustices, violence, and destruction may prevail, or perhaps by those who suggest that this is the best of possible worlds (which may also serve as a legitimation of the status quo). Note that the judgment that something ought not to have happened is no longer a problem of moral discernment only since a clear distinction between good and evil is implied. It soon becomes a metaphysical if not a theological problem: Why did it happen? How did it come to this? This applies both to natural evil (symbolized by the Lisbon earthquake in 1755) and to social evil (symbolized by Auschwitz).[10] Can the presence of evil be made intelligible? In short, this problem is not unique to theology but characterizes the advent of modernity. Although theological discussions of evil may be discarded by some, the problem of (the persistence of) evil is one that cannot be fathomed merely by Enlightenment. Metaphysics is not readily avoided.

Second, there are equally thorny questions to be addressed around the presumed "universality" of sin. Are all human beings sinners and should sin therefore be "democratized" or is a clear distinction needed here between perpetrators and victims, between sinners and those sinned-against, in the interest of justice? Within Christianity itself there are deep divides between evangelicals and a range of liberation, feminist, black, postcolonial, and indigenous theologies. The same divides are present in discourse on climate justice: if there is a need for a "common but differentiated responsibility" to address carbon mitigation, some place the emphasis on the common responsibility (we are all together in the same boat) while others on the differentiated responsibility (focusing on climate justice and climate debt with regard to resource plundering, unfair trade, environmental damage, and waste disposal). It is therefore not surprising that theological discourse on climate change elicits reflection on liberation, remorse, confessing guilt,[11] forgiveness, reconciliation,[12] justification, and justice.[13]

WHOSE ANTHROPOCENE? WHAT DIAGNOSIS?

The second institutional and personal context that shaped this contribution is the opportunity provided by a fellowship on "The Ethics of the Anthropocene" hosted at the Institute for Environmental Studies at the Vrije Universiteit Amsterdam. I framed my project proposal for this fellowship in terms of two questions: "Whose Anthropocene? What Diagnosis?"

Whose Anthropocene is it?[14] Discourse on the Anthropocene, also in the humanities, is heavily dominated by scholars from the global North. Indeed, very few scholars in the fields of ethics, religion, or theology are engaging with the notion of the Anthropocene from the perspective of the global South, that is, scholars who are situated within the global South with interlocutors coming from the global South.[15] In a recent review of 1,209 journal articles published between 2002 and 2016, Jens Marquardt (Stockholm University) comes to the conclusion that "ideas, worldviews, and concepts from the Global South are rarely recognized in the Anthropocene discourse despite the concept's global aspirations."[16] While some are concerned about climate change and ecological destruction, others focus on issues of justice related to economic inequalities, violent conflict, migration, and health concerns. What concerns and interests are at stake here? Whose Anthropocene is it really? How, then, is discourse on the Anthropocene constructed across North/South divides, however these may be construed—West/East, North/South, industrialized/industrializing, developed/developing, Christian/Muslim cultures, center/periphery? As Carmen Gonzalez observes, "In almost every area of environmental concern, North-South negotiations have featured a deep and growing chasm between the call by some Northern states for collective action to protect the environment and the South's demand for social and economic justice."[17] And what about South-South tensions over agricultural land, markets, and natural resources?

This has implications for the second question, namely, regarding an appropriate diagnosis. There is something odd about the very naming of the Anthropocene. Overtly, many would suggest that it demonstrates the ecological impact of anthropocentrism as an ideological disposition that has become predominant at least since the rise of modernity and is epitomized by industrialized capitalism. Yet, the very term also seems to express such anthropocentrism by highlighting as it does the exceptional role of humans in the refashioning of life on Earth.[18] As a result, there is considerable controversy regarding the generalizations embedded in the term "Anthropocene." How is this *anthropos* to be understood across divides of race, gender, class, caste, culture, and geographical location? Put differently: are the root causes of anthropogenic climate change related to the human species as such and in general ("species thinking"), or is the pedigree of this *anthropos* rather European (Greek?), wealthy, white, and male? Do the planetary pretensions of this *anthropos* hide its histories of migration, subjugation, and further migration? How does this shape the diagnosis of what has gone wrong? Put bluntly for the moment, is the underlying problem anthropocentrism or capitalism, inadvertent consumerism or structural violence?

I will return to these questions in the discussion below but have to acknowledge my own situatedness here as a (formerly classified) "white," heterosexual, male, middle- to upper-middle-class Christian theologian, located at a historically "black" institution (UWC), visiting a Dutch university (the VU Amsterdam) with neo-Calvinist roots. This is an intellectual tradition that has played a devastating role in the theological legitimization of apartheid in the South African context, even though, in the Dutch context, it expressed concern for the "kleine luyden" (little people). There is now a Memorandum of Understanding in place between these two universities and cooperation at many levels regarding postgraduate supervision and research.

A hermeneutics of suspicion, for example, one associated with postcolonial and decolonial critiques, is clearly required here. Nevertheless, such a personal and institutional situatedness may also help to recognize the fault-lines that are present in discourse on the Anthropocene in the humanities in the global North. The argument of this contribution is that a retrieval of conceptual tools derived from the theological critique against apartheid in South Africa may offer one authentic contribution from within the global South to multidisciplinary discourse on what is wrong with the world in an age that many describe as "the Anthropocene." Given the situatedness as outlined above, I believe that this is the kind of contribution emerging from the global South that I can personally offer with some integrity.

Although my focus is primarily on secular literature on the Anthropocene, this is best understood as a contribution to Christian ecotheology. This is especially evident from the two toolkits employed (see chapter 2), from the categories of idolatry and heresy discussed below and then from the postscript where I put some of my theological cards on the table. I nevertheless trust that contribution will also appeal to scholars in other disciplines who are interested in an adequate diagnostic of what is *ultimately* wrong with the world.

NOTES

1. For the main outcome of this project, see Ernst M. Conradie, Sigurd Bergmann, Celia E. Deane-Drummond and Denis Edwards (eds), *Christian Faith and the Earth: Current Paths and Emerging Horizons in Ecotheology* (London: T&T Clark, 2014).

2. I do take note of Timothy Morton's observation that "the concept *world* is no longer operational." He regards the socially constructed term "world" as rather anthropogenic and is content to accept "the end of the world." He argues that this end already happened and even offers precise dates for that, that is, 1784 (the patenting of the steam engine) and 1945 (the tests and deployment of nuclear bombs). Instead, he suggests that the Anthropocene announced the encroachment of what he calls hyperobjects (things that are massively distributed in space and time because they out scale humans), given his commitment to object-oriented ontology. Indeed, "The 'world' as the significant totality of what is the case is strictly unimaginable, and for a good reason: it does not exist." See Timothy Morton, *Hyperobjects: Philosophy and Ecology after the End of the World* (Minneapolis: University of Minnesota Press, 2013), 1, 6, 99–133 (108). In this text I use "world" deliberately in a colloquial way, as the world around us (*Umwelt*). This does allow for a Heideggerian line of inquiry (which Morton adopts) but one that is indeed in need of critique or, as I am suggesting, an adequate diagnosis.

3. This is the argument that I developed in Ernst M. Conradie, *Redeeming Sin? Social Diagnostics amid Ecological Destruction* (Lanham: Lexington Books, 2017).

4. See Ernst M. Conradie, "Penultimate Perspectives on the Root Causes of Environmental Destruction in Africa," *Scriptura* 115 (2016): 1–19.

5. See Ernst M. Conradie, "What Diagnosis? Which Remedy? Critical reflections on the Diagnostic Overview of South Africa's National Planning Commission," *Scriptura* 117 (2018): 1–21.

6. I have explored these issues in *Redeeming Sin?* (2017) and in essays with various conversation partners. See Ernst M. Conradie, "John Haught on Original Sin: A Conversation," *HTS Theological Studies* 72:4 (2016): 1–10; "The Christian Faith and Evolution: An Evolving, Unresolved Debate," *Verbum et Ecclesia* 39:1 (2018): 1–13; "On Social Evil and Natural Evil: In Conversation with Christopher Southgate," *Zygon: Journal of Religion and Science* 53:3 (2018): 752–65; "Do Only Humans Sin? In Conversation with Frans de Waal," in *Are We Special? Science and Theology Questioning Human Uniqueness*, edited by Michael Fuller and Dirk Evers (Heidelberg: Springer, 2017), 117–35.

7. See also Ernst M. Conradie, "Divine Election and Migration: The Worst Possible Way to Address the Predicament of Refugees?" *Stellenbosch Theological Journal* 2:1 (2016): 131–48.

8. Again, I have explored the underlying issues in various publications, including Ernst M. Conradie, "Eat and/or be Eaten: The Evolutionary Roots of Violence?" *Scriptura* 114 (2015): 1–22; "What Do We Do when We Eat? Part 1: An Inconclusive Inquiry," *Scriptura* 115 (2016): 1–17; "What Do We Do when We Eat? Part 2: A Theological Inquiry," *Scriptura* 115 (2016): 1–19; and "Could Eating other Creatures be a Way of Discovering their Intrinsic Value?" *Journal of Theology for Southern Africa* 164 (2019): 26–39.

9. See the widely appreciated work on sin by Cornelius Plantinga, *Not the Way it's Supposed to Be: A Breviary of Sin* (Grand Rapids: Eerdmans, 1995).

10. See Susan Neiman, *Evil in Modern Thought: An Alternative History of Philosophy* (Princeton and Oxford: Princeton University Press, 2002), 5. Her book offers a sustained reflection on how the relationship between natural and social evil has characterized modern philosophy. She also recognizes the secularizing tendency in this debate: "The distinction between natural and moral evil began as a debate about how much of the world's misery was God's fault, how much of it ours. Once God was overcome as a human projection, the distinction itself must be overturned" (107). Neiman adds: "Humankind lost faith in the world at Lisbon, and faith in itself at Auschwitz" (250). Lisbon signaled that traditional theodicies were hopeless, Auschwitz that every replacement fared no better (281).

11. See my earlier essay, Ernst M. Conradie, "Confessing Guilt in the Context of Climate Change: Some South African Perspectives," in *Ecological Awareness: Exploring Religion, Ethics and Aesthetics*, edited by Sigurd Bergmann and Heather Eaton (Berlin: LIT Verlag, 2011), 77–96.

12. For some conceptual groundwork, see Ernst M. Conradie (ed.), *Reconciliation as a Guiding Vision for South Africa?* (Stellenbosch: SUN Press, 2013).

13. See my contributions in this regard, including Ernst M. Conradie, "Climate Justice, Food Security . . . and God: Some Reflections from the Perspective of Ecotheology," in *Climate Justice and Food Security: Theological Education and Christian Leadership Development*, edited by Dietrich Werner and Elizabeth Jeglitzka (Geneva: Globethics.net, 2016), 109–34; "To Cover the Many Sins of Galamsey Mining," *Missionalia* 46:1 (2018): 109–30; also *Om Reg te Stel: Oor Regverdiging én Geregtigheid* (Wellington: Bybelmedia, 2018).

14. I am not the only one to raise this question. I found a course offered by Renzo Taddei entitled "Whose Anthropocene? Global South Perspectives on Environmental Crisis" offered at the John Hope Franklin Center for Interdisciplinary and International Studies, Duke University in 2018.

15. There are many scholars from the global North speaking on the Anthropocene on behalf of the global South, some scholars from the global South writing about the Anthropocene who are now located in the global North (e.g., Dipesh Chakrabarty, Kathryn Yusoff), but very few who are actually located in the global South who have interlocutors and implied readers from the global South. One exception in the South African context is Achille Mbembe. See his paper entitled "Decolonizing Knowledge and the Question of the Archive," https://wiser.wits.ac.za/system/files/Achille%20Mbembe%20-%20Decolonizing%20Knowledge%20and%20the%20Question%20of%20the%20Archive.pdf (accessed November 30, 2018).

16. See Jens Marquardt, "Worlds Apart? The Global South and the Anthropocene," in *The Anthropocene Debate and Political Science*, edited by Thomas Hickmann and Lena Partzsch (London: Routledge, 2018), 200–18. Marquardt notes that the global South is framed in discourse on the Anthropocene in terms of especially four key narratives, that is, "helping the global South," "making the Global South accountable," "guiding the Global South," and "assimilating the Global South." Each of these is deeply problematic and subject to critique. Marquardt comments that "the Global South is primarily framed as vulnerable to the effects of anthropogenic environmental

effects, passive in terms of its own problem-solving capacity, and dependent on technological, institutional and social innovations from the Global North. Instead of providing alternatives to the growth-based development model, voices from the Global South are marginalised as they are in related academic debates about topics like global environmental governance . . . or climate justice" (211).

17. Carmen G. Gonzalez, "Bridging the North-South Divide: International Environmental Law in the Anthropocene," *Environmental Law Review* 32:2 (2015): 407–34 (409).

18. Gissli Palsson et al., "Reconceptualizing the 'Anthropos' in the Anthropocene: Integrating the Social Sciences and Humanities in Global Environmental Change Research," *Environmental Science and Policy* 28 (2013): 3–13 (7).

Chapter 1

A Role for Social Diagnostics?

DISCOURSE ON THE ANTHROPOCENE IN THE "NATURAL" SCIENCES

The term Anthropocene is derived from the Greek terms *anthropos* (human) and *kainós* (new, recent). It is best used as the proposed marker of a geological epoch in which the significant recent impact of the human species on the geological layers of the Earth is recognized. Here is one authoritative definition: "The Anthropocene is a proposed new geological epoch based on the observation that human impacts on essential planetary processes have become so profound that they have driven the Earth out of the Holocene epoch in which agriculture, sedentary communities, and eventually, socially and technologically complex human societies developed."[1] Arguably, such impact will remain visible in layers of rock millions of years from now, for example, in terms of the residues of nuclear radioactivity, plastics, and the disappearance of multiple species from the fossil record.

Beyond such stratigraphic evidence of human impact, one may argue from the perspective of Earth system science that the Anthropocene indicates a shift in the way the integrated planetary system functions, given the interplay between solar radiation; the impact of other planets, comets, and asteroids; the Earth's orbital movements; and the lithosphere, the hydrosphere, the atmosphere, the biosphere, and the noosphere, including the technosphere. The field of Earth system science emerged over the past few decades based on the recognition that "The Earth System behaves as a single, self-regulating system comprised of physical, chemical, biological and human components."[2] It is a single, dynamic system, not a collection of ecosystems. This signals a shift from earlier forms of systems thinking to consider the Earth system as a whole, including the human species. It is within this context that the term Anthropocene emerged and is best understood.[3] The Anthropocene cannot be reduced to human impact on ecosystems (i.e., environmental footprint) or even on the Earth as a whole, but is best understood as human impact on the interaction between the various spheres within the Earth system.

As the landmark "2001 Amsterdam Declaration on Earth System Science," issued by the International Geosphere-Biosphere Programme (IGBP) that focuses on Earth system science, observes, "The interactions and feedbacks between the component parts are complex and exhibit multiscale temporal and spatial variability." This statement added the following:

> Human activities are significantly influencing Earth's environment in many ways in addition to greenhouse gas emissions and climate change. Anthropogenic changes to Earth's land surface, oceans, coasts and atmosphere and to biological diversity, the water cycle and biogeochemical cycles are clearly identifiable beyond natural variability. They are equal to some of the great forces of nature in their extent and impact. Many are accelerating. Global change is real and is happening now.
>
> Earth System dynamics are characterized by critical thresholds and abrupt changes. Human activities could inadvertently trigger such changes with severe consequences for Earth's environment and inhabitants. The Earth System has operated in different states over the last half million years, with abrupt transitions (a decade or less) sometimes occurring between them. Human activities have the potential to switch the Earth System to alternative modes of operation that may prove irreversible and less hospitable to humans and other life. The probability of a human-driven abrupt change in Earth's environment has yet to be quantified but is not negligible.
>
> In terms of some key environmental parameters, the Earth System has moved well outside the range of the natural variability exhibited over the last half million years at least. The nature of changes now occurring simultaneously in the Earth System, their magnitudes, and rates of change are unprecedented. The Earth is currently operating in a no-analogue state.[4]

In the opening statement of the significant 2004 IGBP report, the nature of *global* change is captured in the following way:

> Crucial to the emergence of this perspective has been the dawning awareness of two aspects of Earth System functioning. First, that the Earth itself is a single system within which the biosphere is an active, essential component. Secondly, that human activities are now so pervasive and profound in their consequences that they affect the Earth at a global scale in complex, interactive and apparently accelerating ways; humans now have the capacity to alter the Earth System in

ways that threaten the very processes and components, both biotic and abiotic, upon which the human species depends.[5]

The same study also explains what is new in the Anthropocene compared to earlier environmental impact.

> What makes this moment different from those in the past, however, is the global character of the problem—human activity anywhere affects everywhere through systemic change. Past change largely affected the condition of the environment at specific locations. Change now affects global-scale biogeochemical flows, threatening the functioning of the Earth System as a whole in ways not yet fully understood and leading to increasing chances of surprises.[6]

In the so-called natural sciences[7] there remain numerous debates on the adequacy of the term Anthropocene, on the appropriate dating, and on whether this is best regarded in terms of the geological timescale as a boundary event or a transition, an age, epoch (the Anthropocene to succeed the Holocene and Pleistocene), a period (to follow the Quaternary), an era (the Anthropozoic to succeed the Cenozoic), or even an aeon. Since a discussion of these debates falls well beyond my competence, I will only offer a few cursory remarks in this regard.

Like any other species, humans have always changed the environments in which they lived, and this has an impact on larger systems. So what is new about the Anthropocene?[8] From the perspective of the natural sciences, questions around dating and stratigraphic markers for such dating correlate with questions about what constitutes the Anthropocene.[9] The Anthropocene is sometimes dated back to the wave of extinctions of Pleistocene mega-fauna or the agricultural revolution and is thus understood as a continuous process of expanding human impact on the biosphere. Then the Anthropocene becomes, in some sense, natural. Accordingly, as Clive Hamilton puts it, "It is not the product of industrial rapaciousness, an unregulated market, human alienation from nature or excessive faith in technological power; it is merely the result of humans doing what humans are meant to do, that is, use the powers Prometheus gave us to better our lot."[10] If the Anthropocene is associated with the agricultural revolution, this would render "scientific grounding to the desire to defend the status quo against the evidence that the culprit is technoindustrialism's aggressive fossil fuel-driven expansionism, which began at the end of the 18th century."[11]

The Anthropocene may also be dated with the advent of modernity, namely, with reference to the so-called Orbis Spike around 1610—associated with a sudden drop in CO_2 concentrations, probably as a result of mass deaths in the Americas due to epidemics with a resulting change in farming practices and tree growth. If so, it is correlated with the "Columbian exchange" and thus with imperialism and colonialism, if not with capitalism, and the impact this had on all human societies.[12] The implication would be that the geological concept of the Anthropocene may even replace the philosophical notions of "modern" and "modernity."[13] However, such early modern human impact on the atmosphere does not yet imply an impact on the functioning of the Earth system as such.

Alternatively, if the date is associated with the Industrial Revolution (e.g., the commercial use of the steam engine), the "great acceleration" of production and consumption since World War II (favored by several scholars[14]), or better as a two-stage process, then the underlying problem would be associated with the side effects of technological ingenuity,[15] (capitalist) industrial rapacity (in terms of production), and affluenza (in terms of consumption). Then the tendency toward excess can perhaps still be curbed.[16] The argument is that human impact on the functioning of the Earth system as a whole can only be detected with the advent of the Industrial Revolution. It is then the cumulative impact of industrialized civilization, through urbanization, resource extraction, and waste management, that is at stake.[17] Moreover, since the "great acceleration" coincides with the period of decolonization and globalization, this should invite postcolonial and decolonial debate. The atmospheric concentration of CO_2 provides a very useful indicator to track the evolution of the Anthropocene but does not provide a clear stratigraphic starting point due to the functioning of natural sinks of carbon in the oceans and on land.[18]

If the Anthropocene is dated far more precisely (marked by a "golden spike"), namely, to have started on July 16, 1945, with the first nuclear bomb explosion at Alamogordo, New Mexico,[19] then it may seem as if it is the military application of lethal nuclear technology that signals a new dispensation. Then the Anthropocene marks a narrative of elite-driven technology with planet-wide consequences, including the long-term impact of radiation on the victims of Hiroshima, Nagasaki, and subsequent nuclear tests in Oceania.[20]

The challenge for the natural sciences is that the human presence now has to be considered as part of nature—with all the added complexity that the humanities entail. If, in the humanities (with some religious vigor), it was feared that the figure of the human will vanish like an image drawn in the sand, the natural sciences are now confronted with the even more disturbing realization that the human footprint will continue to cast a shadow long after the end of the world (human civilization), albeit not beyond the "vanishing face of Gaia" (life within Earth's biosphere).[21]

Either way, a historical mode of thought is now required in the natural sciences, where a need for geo-stories is evident. Or, as Timothy Morton puts it, "There is no 'pristine,' no Nature, only history. Indeed, . . . Nature is simply reified history."[22] What we call nature (from *natio* = birth) is nothing but "a sum of all births."[23] Moreover, in describing geological and atmospheric changes, natural scientists have become inadvertently, against their own methodological inclinations, prophets,[24] to the extent that their contributions seem to carry apocalyptic overtones.[25] There is another danger, namely, that scientists may proclaim themselves to be the heralded heroes of recognizing the dangers. Accordingly, they construct the narrative of the Anthropocene in such a way that some clever scientists and engineers have to save the day, perhaps with help from the military (!), where politicians, business leaders, and religious leaders have failed to do so.[26]

DISCOURSE ON THE ANTHROPOCENE IN THE HUMANITIES

As many scholars observe, in the Anthropocene a rigid separation between the humanities and the natural sciences is no longer feasible.[27] For some, this implies that the human sciences must play a more significant role in rigorous and more transparent discussions of the Anthropocene Working Group of the International Commission on Stratigraphy and the International Geological Congress where formal decisions are to be made.[28] This implies that geology can no longer be separated from politics and therefore from colonization, slavery, and mineral extraction. Even if the focus of discourse on the Anthropocene from the perspective of the natural sciences must be on Earth systems, the discourse itself needs to be decolonized, as Kathryn Yusoff rightly insists, or else will be subject to the critique of a racialized territorialization of the Earth and branded as "white, colonial geology."[29]

However, this also has implications for how the humanities are construed. In the prescient words of Michel Serres, "Global history enters nature; global nature enters history: this is something utterly new in philosophy."[30] And, for Dipesh Chakrabarty, "The wall of separation between natural and human histories that was erected in early modernity and reinforced in the nineteenth century as the human sciences and their disciplines consolidated themselves has some serious and long-running cracks in it."[31] The underlying problem is that the humanities and social sciences were built on the assumption that humanity has a history of its own—an assumption that is no longer philosophically defensible.[32] While many dualisms have been undermined, often in a reductionist way, the one between nature and culture has proved to be intractable.[33] In the Anthropocene this divide is challenged, almost by definition. Inversely, the messiness of the humanities rubs off on the natural sciences. Geology, for one, is hardly immune, since it is so deeply intertwined with extractive industries in terms of funding, labor, surveying, and technology.[34] Put concisely, what is required is a notion of "nature" in the natural science that includes humanity and a notion of "history" in the humanities that includes the history of life on Earth. This is theoretically perhaps not so difficult to construct but in academic praxis very hard to sustain.

Such a separation also cannot hold between the natural sciences and theology since any recent dating of the Anthropocene typically makes use of the Christian calendar.[35] Indeed, Christianity is deeply entangled with the anthropogenic roots of ecological destruction—most of the countries with high historical carbon emissions were at least nominally Christian by the advent of the Industrial Revolution.[36]

The advent of the Anthropocene confronted the humanities and social sciences with the need to reflect on the impact of this "anthropos" on the planetary systems within which this species emerged. Indeed, what does it mean to be human in the age of humans? A cozy retreat to reflect on the significance of the linguistic, hermeneutical, cultural, sociological, gendered, and spatial "turns" is no longer possible, not least because a discussion of the impact of human behavior cannot be ceded to economics as one branch of the social sciences. John McNeill and Peter Engelke (over-)states this rather bluntly:

> Strangely enough, just as the Great Acceleration was shifting into high gear, academic social scientists and humanists [philosophers, ethicists and theologians too!] chose to retreat from grimy and greasy realities into various never-never lands. They found all manner of discourses worthy of their studied attention, reveling in the linguistic and cultural "turns." But the extinction of species, the incineration of forests, the concentration of CO_2 in the atmosphere—all this seemed unworthy of their powers, interesting only for the discourses it aroused.[37]

Not surprisingly, there has been a virtual explosion of interest in the Anthropocene in a wide range of social sciences and humanities, including philosophy, ethics, religious studies, and theology. However, the term Anthropocene is not (yet) used for a periodization of archaeology or history and hardly ever used in a merely descriptive sense. It may be critiqued for being a homogenizing concept that has become a screen on which other contestations are projected. The emphasis on integrated Earth systems poses a challenge for disciplines in the humanities where a methodological tendency to undermine totalizing grand narratives is evident. This may be appropriate as grand narratives have served the need for an ideological and indeed mythological apparatus of legitimation for the role of Reason, Science, Progress, Growth, or Capitalism. As Peter Sloterdijk observes, despite constructing the course of history on a grand scale, such grand narratives remain provincial: "Because of their incorrigible Eurocentrism, they were in conspiracy with the colonial looting of the world, . . . because they taught salvation history openly or covertly, helped bring profane disaster on a grand scale."[38] Michel Serres's verdict is equally damning: "These . . . intellectuals, believing themselves alone in the world, practiced a universal racism."[39] But, as Clive Hamilton asks, does the arrival of the Anthropocene not provide new grounds for an emerging narrative of humanity as a whole?[40] Or is the assumed anthropos yet another homogenizing strategy?

The Anthropocene has become a highly contested term that almost always includes value-laden connotations. One may suggest that this is an excellent test case for the naturalistic fallacy and hence the need to disentangle descriptive from normative claims.[41] Can one derive an ought from an is? Yes, by all means, especially if what ought to be—and what ought not to be—is already inscribed in what is. Facts and values cannot be distinguished from each other all that rigidly since values rely on a sense of what is real, while all data are theory- and value-laden. Ethics cannot be derived from science but cannot be separated from that either.[42] A prescription for policy-making is often embedded in a factual description of the situation—so that avoiding policy implications requires casting doubt on the facts—as climate change denialists realize very well.[43] However, as the descriptions of the Anthropocene as "good" or "bad," as "sinister or exuberant,"[44] indicate, the actual content of the value judgment is not obvious. Ben Minteer and Stephen Pyne remark that the Anthropocene has become a kind of environmental Rorschach test:

> For some, the notion that humans have become a geological force on the planet—a species able to write its presence in the rocks—is a liberating revelation. It means that we should get on with the business of smart planetary management and get over outdated myths of a separate, pristine wild nature that exists free of human influence. . . . For others, the Anthropocene idea signals the tragic consummation of the destructive human domination of the earth, a last threshold crossed on the march to total ecological despotism.[45]

This implies that the Anthropocene itself (not only the term) is open to moral judgment—in more ways than one—each of which has become contested. There are at least three layers of contestation in discourse on the Anthropocene in the humanities and social sciences that may be mentioned here.

ON NAMING THE ANTHROPOCENE

First, there is an ongoing debate on naming the "Anthropocene" as such. This already implies a twofold judgment: It not only indicates human distinctiveness but also assumes that generalizations about human beings—across divides of geography, race, gender, class, culture, language, and religion—are appropriate, even if this tends to erase histories of patriarchy, white supremacy and (settler) colonialism, dispossession, and oppression.[46] Those who agree that a new marker of geological time is appropriate to distinguish the Holocene from what is following it, given humanity's stratigraphic imprint, remain deeply divided over whether this is best named as the Anthropocene, the proto-Anthropocene,[47] the Capitalocene, the Chthulucene[48] (a poetic rendering of symbiotic relationships), the Eremocine (the age of loneliness following the loss of biodiversity), the Homogocene,[49] the Necrocene (spelling death and extinction), the Oliganthropocene (caused by a small fraction of humans), the Plastocene, the Technocene, or the Thanatocene (ecocide) to list only a few of the available alternatives. Such debates should be welcomed, but only if the basic premise is granted, namely, that (some) humans have become a geophysical force on a planetary scale.[50]

Even if the stratigraphers conclude that the term is not valid to describe a new geological period and even if some scholars in the humanities may object, it seems that the informal use of the term has outstripped the pace of the debate over its status.[51] It is not as if we can do without the word Anthropocene, even if some, like Donna Haraway,[52] would prefer to use it sparingly.

In an important article Andreas Malm and Alf Hornborg question the use of the species category in understanding the Anthropocene narrative and argue that intraspecies inequalities are integral to ecological destruction. They observe that "climate change is denaturalised in one moment—relocated from the sphere of natural causes to that of human activities—only to be renaturalised in the next, when derived from an innate human trait, such as the ability to control fire."[53] They conclude that such species-thinking on climate change is "conducive to mystification and political paralysis" and that it cannot serve as a basis for challenging the vested interests of business as usual.[54]

In her recent *A Billion Black Anthropocenes or None*, Kathryn Yusoff resists alternatives (alter-cenes) to the Anthropocene but insists on a redescription in terms of extractive colonialism as the racialized assemblage from which it emerged.[55] She understands blackness as a "historically and intentionally enacted deformation in the formation of subjectivity, a deformation that presses an inhuman categorization and the inhuman earth into proximity."[56] It is this proximity of black bodies to harm as a result of such intimacy that she refers to as "black Anthropocenes." This usage (deliberately) disables communication between the natural sciences and inhuman geography (Yusoff's discipline) given its critique of "white geology." It also departs from the emphasis on Earth systems that I will follow below.

So whose Anthropocene is it then? In the question raised by Marxist historian Dipesh Chakrabarty,

> . . . does not the talk of species or mankind simply serve to hide the reality of capitalist production and the logic of imperial . . . domination that it fosters? Why should one include the poor of the world—whose carbon footprint is small anyway—by use of such all-inclusive terms as species or mankind when the blame for the current crisis should be squarely laid at the door of the rich nations in the first place and of the richer classes in the poorer ones?[57]

Such resistance against "species-thinking," lumping all human beings under the same category, is understandable since this is exactly the opposite of the direction in which the humanities have been going, notably in the North Atlantic context, where the emphasis has been on difference and otherness, not on identity or any "universal" categories. Strangely, this does not necessarily apply to the humanities elsewhere in the world—where a common humanity (against racial divides), the inalienable dignity of *all* human beings (against various forms of oppression) and (in Asia) "integral thinking" (against Western dualities) have typically been stressed alongside an emphasis on difference. Difference is important in order to resist hegemony and domination but it can inadvertently backfire to allow self-marginalization. Remarkably, the universality of the genetic code nevertheless gives rise to a plurality of species, specimens, languages, and cultures where no one resembles even its closest relatives.[58] As Michel Serres observes, a recognition of the universal does not necessarily give rise to uniformity, only "stupid force" can enforce hegemony, and this is always unsuccessful. The singularity of the big bang generates, gives rise to, makes, all the difference. While globalization opposes difference, the universal encourages it.[59] "Everything dies from uniformity."[60]

Let me state my position in this regard as briefly as possible. I recognize the weighty arguments raised by Jason Moore and many others that it is inappropriate to name an undifferentiated "anthropos" as the distinctive feature of such a new geological epoch. This *anthropos* is predominantly white, male, consumer class, and of recent European origin. The history of imperialism, colonialism, and capitalism was arguably built on excluding most humans from humanity by classifying various categories of people as part of nature but not "society."[61] In this way, their labor could be appropriated in the same way that natural resources are treated as freely available for exploitation.[62] All the global divides that allow for domination in the name of differences of race, gender, class, culture, language, sexual orientation, and so forth are obviously relevant here. This also applies to divides in carbon footprint—in terms of historic carbon emissions between countries, current per capita carbon footprint between countries and between the so-called consumer class[63] and the poor within countries. Although all make use of facilities in the public domain (e.g., roads and their carbon footprint), this differs significantly within a population and between populations in different contexts. Moreover, there are vast differences between the consumption of products (and their embodied energy) and more direct forms of carbon emissions through the use of electricity and for transport.

It is therefore necessary to trace the complex history behind the advent of what is called the "Anthropocene," in which Judeo-Christianity, patriarchy, modernity (Descartes and Kant), the rise of corporations and capitalism, the North Atlantic slave trade, imperialism and colonialism, technology and industry each played a role. Moreover, by differentiating the *anthropos* from other species and the rest of nature, the term Anthropocene seems to maintain the devastating binary between humanity and nature (if not a Cartesian dualism), even as it disturbs the distinction between the natural and the social sciences. The eco-Marxist emphasis on world-systems theory (following Immanuel Wallerstein and others) may indeed be fruitfully extended toward Earth system theory.[64] There is surely a link between globalized capitalism and global warming. That link is obviously the globe that seems too small to fulfill the aspirations of an overheated global economy.[65] Why, then, should one include the world's poor (those whose carbon footprint is small) in all-inclusive terms such as species or humankind when the blame for a changing climate should be placed squarely on imperialist countries in the North Atlantic and on the consumer class elsewhere?[66] It is unfair to blame ordinary people while the ecological crimes of big corporations (and their shareholders) are left unpunished.[67]

Nevertheless, precisely given such complexity, the term "Capitalocene" risks simplifying the roots of environmental destruction to the level of monocausality (capital vs. labor), while the term Anthropocene is more open to a wider range of root causes, including industrialized communism[68] and especially imperialism, understood as structures of dominance maintained by the world's most powerful states.[69] To focus on capitalism only may reveal a Eurocentric bias in discourse on the Anthropocene.[70] The current crisis cannot be reduced to a story of capitalism.[71] For one thing, the periods of capitalism and the Anthropocene do not overlap since capitalism was there long before its geological impact became evident (which is crucial for gathering scientific evidence), while the Anthropocene will continue long after the demise of capitalism.[72] If I am not mistaken, some of the ire with which the name Anthropocene is rejected is prompted by advocates of the "good Anthropocene" who regard it as a human achievement to be celebrated, while the side effects can then be addressed through human ingenuity. Such a use of the term Anthropocene is indeed arrogantly anthropocentric, but the term is not necessarily used in that way in current discourse.[73]

My sense is that Clive Hamilton captures the underlying problem, namely, that any claims for exceptionalism (that some groups of humans are not responsible) are facile. He acknowledges that the model of capitalist industrialization pursued by China, India, Brazil, and other countries of the South is European in origin so that the *anthropos* in the name Anthropocene hides its European pedigree.[74] It is the American dream that has purchased the Anthropocene nightmare.[75] The totalizing forces of globalization have yielded the Anthropocene. Nevertheless, even if we ascribe ecological destruction to the hegemonic power of capital, the question remains why others have so willingly collaborated with it (through the fetish of economic growth, faith in technological fixes, short-term political solutions, and the innate self-absorption of consumerism), even if their futures are being destroyed?[76] In any case, even if all humans are not responsible for bringing about the Anthropocene, every human is destined to live within it.[77] This unification of humanity under the ubiquitous banner of the shopping mall (as buyers, workers, or beggars) may be unwelcome but is nevertheless unmistakable.

Hamilton, therefore, argues that we need to locate the emergence of industrial capitalism within the history of the species in order to avoid seeing it as a contingent outcome of technological breakthroughs and colonial exploitation.[78] Humans are indeed not equally responsible for bringing on the Anthropocene, but this responsibility is increasingly shared between the North and the South that includes the southern BRICS countries, namely, China, India, Brazil, and South Africa. The main reason why it is appropriate to attach the new epoch to an undifferentiated *anthropos* is that countries still included under the global South will cause as much, if not more damage to the global climate system as the North as it copies the northern model of environmental harm in the pursuit of poverty alleviation and affluence.[79] Moreover, it may become a facile excuse for countries from the global South to cite issues related to climate debt and an unfair distribution of climate in order to shirk their own responsibilities.[80] Hamilton insists that

> it is time to abandon the "dependency theory" developed in the 1960s and 1970s to explain continued subservience of the South after decolonization, and later elaborated into world-system analysis (according to which the globe is organized around an international division of labor and capital in which the "core" or rich countries exploits the "periphery" of poor ones). . . . The understanding of the globe that emerged in the 1950s and 1960s as one divided between North and South has been overturned by history, and the *rapture* of the Anthropocene arrives to challenge social science to build new theories.[81]

Hamilton concludes, "From an Earth System viewpoint, there are on Earth no divisions between North and South or between nations, cultures, genders, and races. There are only humans with more or less power to disturb it."[82] The history of capital is mixed with the history of the species.[83] From the perspective of biodiversity, it is the domination of one species over others that destabilizes the system. The tyranny of the Earth System will overrule the plurality of local stories and cultures.[84] The question is not whether the implied *anthropos* is inclusive or representative of all human beings, or whether all are culpable; instead the point is that the impact of the powerful on Earth systems dialectically reverts to encapsulate all human beings, and to impose the guilt of the not so few on the many.

It may be helpful to contextualize the Anthropocene by pointing to the diversity of ways in which bioregions, countries, cities, and sectors of the population engage with its multiple aspects.[85] Likewise, it is crucial to expose assumptions about Anthropocene universality as disguising subjugation in terms of gender and especially race. If humanity has become a geological force of nature, this is racialized through slave labor on mines, plantations, and factories.[86] The strength of the humanities has been to resist claims for universality and to undermine any hegemony by demonstrating not only an irreducible plurality but also the multiple inequalities and forms of domination in the name of such differences. However, such calls for locality and difference seem to miss the emphasis on integrated Earth systems that is definitive of the Anthropocene. This may be a bitter pill to swallow for the humanities, but this is exactly why the divide between the so-called natural and the human sciences is so troubling.

It would in my view therefore not do to change the name of the Anthropocene to the Capitalocene (as another example of Western linguistic imperialism) since this will not exempt the poor and vulnerable from its devastating impact.[87] And, as Dipesh Chakrabarty adds, the term Anthropocene "now refers more to (mostly human-driven) changes to Earth System as a whole and less about moral culpability of humans (or some humans) in causing them."[88] In reviewing the intertwined historical roots of a globalized carbon economy, both West and East, Amitav Ghosh acknowledges the truth in the bureaucratic formula of a "common but differentiated responsibility." He accepts that anthropogenic climate change is the consequence of the very existence of human beings as a species:

> Although different groups of people have contributed to it in vastly different measure, global warming is ultimately the product of the totality of human actions over time. Every human being who has ever lived has played a part in making us the dominant species on this planet, and in this sense every human being, past and present, has contributed to the present cycle of climate change.[89]

It remains a question whether a change in terminology alone would do, although nicknames should be welcome. The "Capitalocene" does signal how global inequality and capitalism's dependence on cheap natural resources have produced the current predicament.[90] Nomenclature does political work; a single word can create a discursive infrastructure for political change. However, as Gabrielle Hecht argues, "words have such power only if they're widely adopted, and it's hard to imagine geologists or climate scientists embracing alternatives. The political leverage of the Anthropocene concept lies precisely in its analytic potential to bring together researchers across the natural, social and human sciences—as well as the arts—in order to better understand the complex dynamics that put our species at risk."[91]

In accepting the term Anthropocene, for better or for worse, it may still be true that confronting capitalism in its strongholds is the key struggle that can address the more planetary wide problem of anthropogenic impact on the Earth system. In other words, the common-sense notion that we first have to tackle the ecological problem for the sake of survival may be misleading.[92] As I will suggest below, the ecological problem is not really of an ecological nature but economic, political, cultural, and indeed spiritual in scope.

As an aside, I may note that this debate is structurally analogous to a debate in Christian theology on the universality of sin. Following Augustine and Luther, evangelical theologies have typically maintained that all human beings are sinners, that only human beings are fallen, and that sins cannot be graded. By contrast, several contemporary theological movements have emphasized a clear distinction between those who are sinners and those who are sinned against. To underplay such a distinction is to undermine the need for justice and to hold the victims of history responsible for their own oppression. It is, for example, obscene to argue that both rapists and rape victims are sinners. However, there is also a need to recognize that the victims of yesterday may become the perpetrators of tomorrow. Likewise, in climate change parlance those who are retrospectively guilty (Europe and the United States), may soon be joined by those who are prospectively guilty (China and later India).[93]

I have argued elsewhere that the proportionality of guilt does not exclude a recognition of the universality (or better the planetary scope of the impact) of sin.[94] We are all (all living creatures) in this mess (the Anthropocene) together. None of us chose this mess, at least not deliberately (which offers no excuse anyway). We (all humans) contributed to this mess in one way or another, as moral agents or as moral patients (on whose behalf others act), but not equally so. We (will) all suffer under it but again not equally so. None of us can avoid the mess either.[95] Augustinians will add that we cannot escape from this mess either, at least not by our own human powers—derived from science and technology perhaps, or from activist struggles to "save the earth."

From a theological point of view, I would prefer to call the Anthropocene the "Hamartiocene," that is, the age in which the global impact of human sin has become evident. Since such nomenclature will not be widely adopted, I will continue using the term Anthropocene, albeit with the many reservations that have been raised in such current debates.

What is then the short answer to the question whose Anthropocene it is? I would say the consumer class globally. The longer answer? Those who aspire to become part of the consumer class, preferably soon. Admittedly, this is a cultural answer that does not take account of the complexities of politics, business and industry, science or philosophy, but there is some virtue in the simplicity of this take on the Anthropocene.

VALUE JUDGMENTS CONCERNING THE ANTHROPOCENE

If bacteria and other forms of life have altered Earth systems for billions of years, human impact is not by itself unique. What is qualitatively new is that this is a species that is aware of doing so and is accountable for its actions, given moral agency.[96] It is therefore not surprising that the "Anthropocene" is a value-laden term. If so, a second layer of contestation in discourse on the Anthropocene has to do with the implied moral judgment. Some may be averse to any such moral judgment given the many contestations that this entails. From a "purely scientific" (read positivist) point of view, the Anthropocene is neither good nor bad, it is a merely a descriptive term, "a fact supported by overwhelming evidence."[97] Yet, it is equally obvious to say that biodiversity is better than monocultures, that flourishing is healthy and being poisoned is, well, toxic. Or that a vibrant ecosystem is good and mass extinction bad, except on the very long term.[98] It is therefore not surprising that a recent article states that "Scientists have a *moral* obligation to clearly warn humanity of any catastrophic threat and to 'tell it like it is.'" With some 11,000 signatories from around the world these scientists state "clearly and unequivocally that planet Earth is facing a climate emergency."[99]

A purist position on avoiding the naturalistic fallacy is therefore hardly possible. At the same time, the Anthropocene illustrates the trouble that arises when what ought to be (or not to be) becomes embedded in what is. It conflates moral evil and natural evil so that any distinction becomes harder to make because the cumulative impact of moral evil has become a (geological) force of nature. A clear distinction between what is and what ought to be is needed but is harder than ever before. There are some ironies here. While science thrives on a healthy skepticism and demands room for contrarian views, climate denialists are better understood as dogmatists who espouse views despite evidence to the contrary.[100] Inversely the IPCC was awarded the Nobel peace prize rather the prize for physics or chemistry![101] Bruno Latour notices the following:

> We are now attributing to natural history the terms of human history—tipping points, acceleration, revolution—and that to speak of human history we are using the words inertia, hysteresis, path dependency, as if humans had taken on the aspect of a passive and immutable nature in order to explain why they are doing nothing against the threat . . . the old distance between background and foreground has faded away: it is human history that appears cold and natural history that is taking on a frenzied aspect.[102]

Moreover, the Anthropocene seems to confirm a radical constructivist position. Not only human knowledge, scientific enterprises, social institutions, political economies, and gender inequalities are socially constructed. Even the chemical composition of the atmosphere is the product of the social construction of reality, embedded in power relationships. As Latour observes, "What literary critic would have extended the principles for deconstructing texts to the sedimentary strata revealing in all the deltas of the planet the irrefutable traces of erosion caused by humans?"[103] If the distinction between facts and values, light and enlightenment, becomes so radically questioned, it may even be appropriate to consider not only a "genealogy of morality" (Nietzsche) but also a "geology of morality" (Deleuze).[104] Yet, conflating science with ethics (or religion), the state of nature with the nature of the state (Hobbes!)[105] remains as dangerous as ever before, even if separation is no longer feasible. If we are tampering with the weather, the distinction between natural evil and social evil is no longer clear-cut. This is not merely a theoretical problem, at least not for insurance companies trying to assess what may or may not count as an "act of God."[106]

The Anthropocene is clearly not an amoral term, but "moral" or "immoral" can hardly describe a whole civilization, not to mention a geological period. To frame the Anthropocene as a matter of ethics seems to trivialize an event that is without parallel, one what is, in fact, Earth-shattering.[107] Nevertheless, the question remains whether it is possible to speak of the "good Anthropocene," something to be welcomed and embraced rather than lamented and feared—in the same way that accepting planetary responsibility may be regarded as appropriate.[108] This is how Erle Ellis frames the "good Anthropocene" in an article in the journal of the Breakthrough Institute:

> The Earth we have inherited from our ancestors is now our responsibility. It is not natural limits that will determine whether this planet will sustain a robust measure of its evolutionary inheritance into the future. Our powers may yet

> exceed our ability to manage them, but there is no alternative except to shoulder the mantle of planetary stewardship. A good, or at least a better, Anthropocene is within our grasp. Creating that future will mean going beyond fears of transgressing natural limits and nostalgic hopes of returning to some pastoral or pristine era. Most of all, we must not see the Anthropocene as a crisis, but as the beginning of a new geological epoch ripe with human-directed opportunity.[109]

In a later contribution, Ellis asks whether this is "a story of senseless destruction or a story of awakening and redemption"?[110] He argues that, on average, current generations live longer, healthier, less violent lives, are better educated, and have access to opportunities unimagined by their ancestors.[111] Ellis nevertheless adopts a more cautious tone and answers the question by inviting a range of Anthropocene narratives because no single account can suffice.[112] He speaks of Anthropocenes in the plural because he believes that "prospects for anthropocenes much better than the one we are now creating are very real."[113] As evidence he points to a number of success stories in dealing with ozone depletion, banning DDT, combating acid rain, and protecting wildlife. We therefore should not abandon the process of modernization but should embrace human power, that is, technology. If so, the Anthropocene is both a challenge and an opportunity.[114]

Others suggest that the advent of the Anthropocene is a catastrophe that spells the reversal of the "unfinished project of modernity" (Habermas), that is about to fall to ruins, that may bring the end of human civilization, of the human species and is indicative of a "sixth extinction."[115] They resist attempts to revise the meaning of the term Anthropocene by suggesting that whatever limits, thresholds, and tipping points the Earth system may pose, human ingenuity and technology will be able to transcend them.

Is the Anthropocene then an unmitigated disaster, a topic that can only be addressed in fear and shame, or is our ability to alter the atmosphere a significant achievement for which we now need to take responsibility by addressing the side effects? Should the positive outcomes of Enlightenment and industrialization be discarded? The advocates of geoengineering are best described as eco-modernist in orientation given the emphasis on humanity's creative power and ability to take its destiny into its own hands. As Clive Hamilton, Christophe Bonneuil and François Gemene comment on geoengineering strategies in the editorial of their significant volume *The Anthropocene and the Global Environmental Crisis*,

> These are the kinds of strategies we can expect to see mobilised to defend and protect the established order as the pillars of modernity crumble. In their belief in human reason, faith in the power of technology and confidence in the nobility of the human quest, they are beautifully modern.[116]

For Simon Dalby, though, the Anthropocene is neither good nor bad but ugly: "The Anthropocene is neither good nor bad but is going to be shaped by a politics that is necessary and probably will be rather ugly given the resistance of the fossil fuel industry in particular to attempts to keep 'rocks' in the ground."[117]

It is fair to say that any attempt to defend the Anthropocene as "good" will come under suspicion in the humanities, especially from the perspective of the global South, as a way of legitimizing neoliberal industrialized capitalism. This also applies to the disciplines of philosophy, ethics, religious studies, and Christian theology from which I will draw in this contribution.

At the same time, especially in the field of theology, one needs to guard against not only a false diagnosis or selective morality but also against any self-righteous diagnostic gaze, as if it is possible to identify the root of all evil. Slavoj Žižek's warning is pertinent here: "Think about a religious fundamentalist who sees signs of sin and corruption everywhere in modern society—is the true evil not his suspicious gaze itself?"[118] Diagnosis is the task of a doctor who is presumably not him- or herself sick. Put differently, does morality demand that we make evil intelligible by accounting for the presence of evil, or does the opposite apply, namely, that morality demands that we do not seek to make evil intelligible?[119] Some would claim that the very attempt to gain an adequate diagnosis is already part of the problem, namely, the folly, the arrogance of a meta-stance, claiming to see a bigger picture that others are too myopic to detect.[120] Indeed, the quest for the knowledge of good and evil is rather close to the idolatrous desire to become divine.[121] Perhaps even the desire for such knowledge, and thus dominion over other things, preempts the death of God.[122]

IN SEARCH OF AN ADEQUATE DIAGNOSIS

A third layer of contestation in discourse on the Anthropcene relates to the obvious debate on an adequate diagnosis of what is wrong in the world, given the Anthropocene as marker. Also those eco-modernists speaking of a "good Anthropocene," even a "great Anthropocene,"[123] are willing to consider the question what is wrong *in* the Anthropocene, if not *with* the Anthropocene. There is no need here to describe the symptoms that manifest themselves. Not only does this require expertise beyond what I can provide but there is also ample literature that offers a description of such symptoms in a publicly accessible way.[124] To stay with medical terminology such a "diagnosis" is typically tied with assessments of appropriate remedies and therapies, a possible prognosis, lifestyle changes that follow from that, and a sense of destiny.[125] Any diagnosis assumes that something has gone wrong that is less than optimal. If so, what went wrong? What was the situation before things went wrong? When did things start going wrong? What is the extent of the problem? Ultimately, the question will emerge: why did that happen? Like with other questions of an ultimate nature, no final answers are to be expected here! As Stephen Gardiner observes, as far as climate change is concerned we are in a stage of initial diagnosis rather than depth diagnosis: "We are still largely at the stage of saying, for example, that climate change is seriously unjust to the global poor, future generations, and nature, rather than at the stage of offering deep analysis of what exactly has gone wrong and what it would take to get it right."[126]

The focus of this contribution will be on this third question, namely, on an appropriate diagnosis of what has gone wrong in (and with) the Anthropocene. The notion of social diagnostics does move beyond the medical or psychological terminology in the sense that what is "wrong" is no longer an amoral condition for which a "remedy" or "therapy" needs to be found but is open to moral discernment. It should be immediately clear, though, that this is an incredibly complex question with shallow, superficial answers readily available. The twentieth century has suffocated us with such diagnoses—national grievances, inadequate development, the oppression of particular groups, the over- or underregulation of the market—each coming with their own prescriptions. As Bruno Latour remarks, any effort to stabilize an ethical judgment by invoking nature will appear as a "scarcely concealed disguise of an ideology," given the vehement disputes over morality.[127] Yet, as Mary Midgley wryly observes, for philosophers there is nothing unprofessional in attending to large issues, in biting off more than one can chew.[128] Or as Peter Sloterdijk comments, "It is characteristic of being human that human beings are presented with tasks that are too difficult for them, without having the option of avoiding them because of their difficulty."[129]

To search for an adequate diagnosis is a task that can only be approached in a multidisciplinary way, involving contributions from philosophy, social theory, anthropology, economics,[130] and politics, to mention some obvious examples. What is required is not only scientific observation, technical expertise, and economic calculus—that privilege a form of rationalism—but also a form of reasoning that are valid in other spheres, "where the weaknesses of humans and their institutions are recognized and the lessons of history absorbed."[131] To assess the Anthropocene entails nothing less than an assessment of the human condition. To "know thyself" has been the core task of philosophy as the haunting inscription at the temple of Apollo in Delphi reminds us. This task is becoming ever harder following the nineteenth-century death of God, the twentieth-century death of the subject, and the twenty-first-century death of history despite Cartesian claims about the cogito and Kantian assumptions about the inalienable dignity of the human subject.[132] As Marx, Nietzsche, and Freud, the masters of suspicion and their followers, recognize, our species cannot be equated with the stories that we tell about ourselves.[133] To dare to give a definitive response (or any response whatsoever) to the question "What is wrong with the world?" would be ludicrous. Not only may there be multiple symptoms and multiple diagnoses (as many as there are philosophers) but also, as I will argue below, different layers of the underlying problem.

The Anthropocene poses an even deeper question, namely, to reflect on what we have become.[134] To know oneself may be the art of living, but the clue may well lie in the art of dying (*ars moriendi*), to prepare for death. As Roy Scranton has it, "The rub now is that we have to learn to die not as individuals, but as a civilization. . . . The greatest challenge we now face is a philosophical one: understanding that this civilization is already dead."[135]

Moreover, even if an accurate diagnosis is possible, this does not imply that a remedy is forthcoming. The disease may be terminal so that treatment may have to be stopped. Nevertheless, there is only one thing that can be worse than offering such a diagnosis and that is not offering a diagnosis, not even attempting to address the question when it is so obviously needed! This contribution will seek not to shy away from such a task.

There is also a different kind of danger associated with seeking an adequate diagnosis, namely, that to dwell on a diagnosis can become a facile excuse not to have to even consider possible remedies. The question what is wrong is the wrong question, some may say. We should focus on the positives in order to find the resources to address the challenges that are there. It is not necessary to perfect the diagnosis and the remedy before implementing intermediary steps on which there are sufficient consensus. As Dale Jamieson wryly observes, it is time to stop letting the perfect be the enemy of the good.[136] Searching for a diagnosis may even be a matter of Schadenfreude or simply an unwillingness to get one's hands dirty while readily shooting down proposed not only solutions but also other diagnoses. This is the danger addressed by calls for an ethics of responsibility, for example, by Hans Jonas. Jeffrey Sachs, following economist Albert Hirschman, identifies three pertinent dangers that constructive attempts to work for social transformation may encounter. First, there is the danger of futility that assumes that social change is impossible because the underlying problem (e.g., rising carbon emissions) is unsolvable within a particular time frame. Second, there is the danger of perversity, namely, the assumption that any attempt to offer a solution will probably just make matters worse (e.g., geoengineering). Third, the danger of jeopardy expresses fears that attempting to solve a problem may divert energy and resources away from more important challenges (e.g., poverty reduction, violent conflict, diseases).[137]

The first two of these dangers may well be relevant for this contribution given the focus on a diagnosis without even addressing possible solutions. As

I will suggest, there is indeed a need to guard against shallow diagnoses and therefore superficial remedies. However, the counterargument may be that the most thorough diagnosis may well emerge only when the availability of a remedy becomes evident. This insight is also embedded in the Christian conviction that a recognition of guilt before God only becomes possible through an encounter with God's forgiving love. As argued by Karl Barth and others, it is not as if we first need to understand what the problem is before we can appreciate how we can be saved from such a problem. Theologically, this gets the order wrong because it assumes that it is possible to gain knowledge of our human needs independent of God's revelation in Jesus Christ.[138] For this reason, I will add a theological postscript to the multilayered diagnosis that is offered here.

NOTES

1. See Will Steffen et al., "Trajectories of the Earth System in the Anthropocene," *Proceedings of the National Academy of Sciences of the United States of America* (August 6, 2018): 1–8 (1).

2. See http://www.igbp.net/about/history/2001amsterdamdeclarationonearthsystemscience.4.1b8ae20512db692f2a680001312.html (accessed January 16, 2019).

3. See Clive Hamilton and Jacques Grinevald, "Was the Anthropocene Anticipated?" *The Anthropocene Review* 2:1 (2015): 59–72. The emergence of the notion of an Anthropocene over the past two decades should not be confused with the question on how the Anthropocene itself is dated (see below).

4. See http://www.igbp.net/about/history/2001amsterdamdeclarationonearthsystemscience.4.1b8ae20512db692f2a680001312.html (accessed January 16, 2019).

5. Will Steffen et al., *Global Change and the Earth System: A Planet under Pressure* (Berlin: Springer, 2004), 1.

6. Ibid., 131.

7. In my view all sciences are "natural" in the sense that humans doing such work form part of nature, while nature as the object studied in such disciplines include the human species. Likewise, all sciences are "human" in the sense that no other species (or angels for that matter) engage in scientific inquiry. In the same vein I have argued elsewhere that all forms of theology are natural theology. See Ernst M. Conradie, "All Theology is Natural Theology: The Hermeneutic Necessity of Natural Theology?" *Ned Geref Teologiese Tydskrif* 52:1&2 (2011): 58–65.

8. I find the emphasis on human systems by Erle Ellis helpful. He distinguishes between basically three such human systems, namely, the Paleolithic system of hunting and gathering communities, the Neolithic agricultural systems, and the Industrial system. Each has had a significant impact on the terrestrial biosphere, but the scale is not the same. The industrial system is novel in its use of fossil fuels, the industrial synthesis of reactive nitrogen to boost agricultural productivity, and genetic

engineering. See Erle C. Ellis, "Anthropogenic Transformation of the Terrestrial Biosphere," *Philosophical Transactions of the Royal Society* 369 (2011): 1010–35.

9. This is often reviewed in discourse on the Anthropocene in the natural sciences. For an authoritative early discussion, see Will Steffen, Jacques Grinevald, Paul Crutzen, and John McNeill, "The Anthropocene: Conceptual and Historical Perspectives," *Philosophical Transactions of the Royal Society* A369:1938 (2011): 842–67.

10. Clive Hamilton, *Earthmasters: The Dawn of the Age of Climate Engineering* (New Haven: Yale University Press, 2013), 202–3.

11. See Clive Hamilton, "The New Environmentalism Will Lead Us to Disaster," *Scientific American*, June 19, 2014, https://www.scientificamerican.com/article/the-new-environmentalism-will-lead-us-to-disaster/ (accessed March 2, 2019).

12. This is the proposal developed by Simon L. Lewis and Mark A. Maslin, *The Human Planet: How We Created the Anthropocene* (Penguin Books, 2018). See especially this: "From a narrative perspective, beginning the Anthropocene with the birth of the modern world tells the story of a new profit-driven mode of living. This new geological epoch is built from slavery and colonialism, enabled by a long-distance financial industry. The human epoch is a story of domination, and the resistance to that domination. The powerful ushered in this new planet-changing epoch, but the reasons for specific expression of how the Anthropocene has evolved are complex. The Earth system has changed over this period due to the relationship between a dominant class of people and the rest: it is the outcome of the dynamics of the actions of the powerful and the desires of billions of people. Those with the power to direct others created the Anthropocene, but its unfolding is the result of the forces of human history, including the impacts of opposing the power that elites wield. This description of the Anthropocene tells of people having the agency to change the world, but our abilities to do so being far from equal" (319–20). They add: "If the Anthropocene is pinned to the Columbian Exchange, the deaths of 50 million people, and the beginnings of the modern world, then it is a deeply uncomfortable story of colonialism, slavery and the birth of a profit-driven capitalist mode of living being intrinsically linked to long-term planetary environmental change. What we do to each other matters, as well as what we do to the environment. And given that nobody meant to transfer diseases that killed tens of millions, it is also a cautionary story: human actions can cause accidents with terrible consequences" (326–27). The deliberate spread of smallpox as a biological weapon in the Americas only exacerbates the argument made by Lewis and Maslin.

13. A proposal welcomed by Bruno Latour, *Facing Gaia: Eight Lectures on the New Climate Regime* (Cambridge: Polity Press, 2017), 116.

14. See, especially, John R. McNeill and Peter Engelke, *The Great Acceleration: An Environmental History of the Anthropocene Since 1945* (Cambridge: The Belknap Press, 2014).

15. Yusoff's critique of this way of telling the story is devastating: "This is the tale of entrepreneurship of a few white men transforming the world with their ingenious creations or of a political economy that is aggressively sutured to the earth's processes via the lifeblood of fossil fuels. So the explorer as hero (Columbus) is

replaced by the inventor as hero (Watt and his engine) in the progress narrative of Man as the agentic center and authority of power, cut with some European genius myth to rarefying the white male subject and his imperial intellectualism." See Kathryn Yusoff, *A Billion Black Anthropocenes or None* (Minneapolis: University of Minnesota Press, 2018), 39.

16. See Hamilton, *Earthmasters*, 204.

17. See Clive Hamilton, Christophe Bonneuil and François Gemene (eds), *The Anthropocene and the Global Environmental Crisis* (London and New York: Routledge, 2015), 2–3.

18. Steffen et al., "The Anthropocene: Conceptual and Historical Perspectives," 848.

19. This is proposed by Jan Zalasiewicz et al., "When did the Anthropocene Begin? A Mid-twentieth Century Boundary Level is Stratigraphically Optimal," *Quaternary International* 383 (2015): 196–203. This proposal may be "stratigraphically optimal" but raises questions how the term Anthropocene is best to be used in the humanities, if at all.

20. See Lewis and Maslin, *The Human Planet*, 327; also Yusoff, *A Billion Black Anthropocenes or None*, 44–48.

21. See Morton, *Hyperobjects*, 195. He refers to Foucault's quip at the end of *The Order of Things*, while "vanish" also appears in the title of James Lovelock's *The Vanishing Face of Gaia: A Final Warning* (New York: Basic Books, 2009).

22. Morton, *Hyperobjects*, 58.

23. Michel Serres, *The Incandescent* (London: Bloomsbury, 2018), 15.

24. See Latour, *Facing Gaia*, 28: "The earth science experts are the ones today who look like over-excited militants of a cause, fanatics, catastrophists, and climate skeptics are the ones assuming the role of stern scientists, who at least do not confuse the way the world is going with the way it ought to go! They even succeeded in appropriating—while reversing its meaning—the fine word 'skeptic.'"

25. There are numerous examples of such almost apocalyptic overtones. See, for one example alongside very many others, Steffen et al., "Trajectories of the Earth System": "In the future, the Earth System could potentially follow many trajectories, often represented by the large range of global temperature rises simulated by climate models. In most analyses, these trajectories are largely driven by the amount of greenhouse gases that human activities have already emitted and will continue to emit into the atmosphere over the rest of this century and beyond—with a presumed quasilinear relationship between cumulative carbon dioxide emissions and global temperature rise. However, here we suggest that biogeophysical feedback processes within the Earth System coupled with direct human degradation of the biosphere may play a more important role than normally assumed, limiting the range of potential future trajectories and potentially eliminating the possibility of the intermediate trajectories. We argue that there is a significant risk that these internal dynamics, especially strong nonlinearities in feedback processes, could become an important or perhaps, even dominant factor in steering the trajectory that the Earth System actually follows over coming centuries" (2). They add: "Avoiding this threshold by creating a stabilized Earth pathway can only be achieved and maintained by a coordinated,

deliberate effort by human societies to manage our relationship with the rest of the Earth System, recognizing that humanity is an integral, interacting component of the system. Humanity is now facing the need for critical decisions and actions that could influence our future for centuries, if not millennia" (7). I will return to the use of apocalyptic imagery in the final chapter.

26. On the grand narrative with scientists as heroes, see Christophe Bonneuil and Jean-Baptiste Fressoz, *The Shock of the Anthropocene: The Earth, History and Us* (London: Verso,2017), 79–83.

27. From the perspective of the humanities, see the first of Dipesh Chakrabarty's four theses: "Anthropogenic explanations of climate change spell the collapse of the age-old humanist distinction between natural history and human history." See Dipesh Chakrabarty, "The Climate of History: Four Theses," *Critical Inquiry* 35:2 (2009): 197–222 (201). From a natural science perspective, see Steffen et al. ("Trajectories of the Earth System," 5): "Human societies and our activities need to be recast as an integral, interacting component of a complex, adaptive Earth System. This framing puts the focus not only on human system dynamics that reduce greenhouse gas emissions but also on those that create or enhance negative feedbacks that reduce the risk that the Earth System will cross a planetary threshold and lock us into a Hothouse Earth pathway."

28. See Erle C. Ellis et al., "Involve Social Scientists in Defining the Anthropocene," *Nature* 540:7632 (2016): 192–93.

29. See Yusoff, *A Billion Black Anthropocenes or None*, 103, 105.

30. Michel Serres, *The Natural Contract*, translated by Elizabeth MaxArthur and William Paulson (Ann Arbor: The University of Michigan Press, 1995), 4.

31. Dipesh Chakrabarty, "Postcolonial Studies and the Challenge of Climate Change," *New Literary History* 43:1 (2012): 1–18 (10).

32. Clive Hamilton, *Defiant Earth: The Fate of Humans in the Anthropocene* (Cambridge: Polity Press, 2017), 8.

33. The most significant attempt in the field of anthropology to address the divide between nature and society may be that of Philippe Descola. See his *Beyond Nature and Culture* (Chicago: University of Chicago Press, 2013).

34. See Yusoff, *A Billion Black Anthropocenes or None*, 13.

35. Steffen et al., "The Anthropocene: Conceptual and Historical Perspectives," 849.

36. This observation forms the point of the departure for the recent *The T&T Clark Handbook on Christian Theology and Climate Change* (London: T&T Clark, 2019), edited by Hilda Koster and myself.

37. McNeill and Engelke, *The Great Acceleration*, 5, 211.

38. Peter Sloterdijk, *In the World Interior of Capital: For a Philosophical Theory of Globalization* (Cambridge: Polity Press, 2014), 4.

39. Serres, *The Incandescent*, 104.

40. Hamilton, *Defiant Earth*, 77.

41. See Maria Antonaccio, "De-moralizing and Re-moralizing the Anthropocene," in *Religion in the Anthropocene*, edited by Celia E. Deane-Drummond, Sigurd Bergmann and Markus Vogt (Eugene: Cascade Books, 2017), 121–37.

42. See J. Baird Callicot, *Thinking Like a Planet: The Land Ethics and the Earth Ethic* (Oxford: Oxford University Press, 2013), 7.

43. See Latour, *Facing Gaia*, 25. His version of the climate denial lobbyists reads: "We do not believe in these facts; they do not suit us; they lead to sacrifices that we do not want to make" (27).

44. See Antonaccio, "De-moralizing and Re-moralizing the Anthropocene," 123. Her aim is to rescue the term Anthropocene from such diverging assessments and, in the process, allow for a concept of nature exterior to the human project.

45. Quoted in Ibid., 125. The reference is to Ben A. Minteer and Stephen J. Pyne, "Writing on Stone, Writing in the Wind," in *After Preservation: Saving Nature in the Age of Humans*, edited by Ben A. Minteer and Stephen J. Pyne (Chicago: University of Chicago Press, 2015), 1–8.

46. One pertinent example is the work of Kathryn Yusoff who traces the color line of discourse on the Anthropocene, with reference to issues of gender and sexual orientation. See her *A Billion Black Anthropocenes or None.*

47. As proposed by David Grinspoon, *Earth in Human Hands: Shaping Our Planet's Future* (New York: Grand Central, 2016), 226. The proto-Anthropocene (characterized by inadvertent random changes) is contrasted with the "mature Anthropocene," that is, "when we fully incorporate our uniquely human powers of imagination, abstraction and foresight into our role as an integral part of the planetary system" (226).

48. See Donna J. Haraway, *Staying with the Trouble: Making Kin in the Chthulucene* (Durham: Duke University Press, 2017).

49. See Timothy Morton, *Dark Ecology: For a Logic of Future Coexistence* (New York: Columbia University Press, 2016), 23.

50. See Ibid., 20.

51. See, especially, Grinspoon, *Earth in Human Hands*, 217.

52. See Donna J. Haraway, "Staying with the Trouble: Anthropocene, Capitalocene, Chthulucene," in *Anthropocene or Capitalocene? Nature, History and the Crisis of Capitalism*, edited by Jason W. Moore (Oakland: PM Press, 2016), 34–76 (51). Latour may be correct in suggesting that "staying with the trouble" means that we should precisely *keep* the word Anthropocene. See Latour, *Facing Gaia*, 100.

53. Andreas Malm and Alf Hornborg, "The Geology of Mankind? A Critique of the Anthropocene Narrative," *The Anthropocene Review* 1:1 (2014): 62–69 (65).

54. Ibid., 67.

55. See Yusoff, *A Billion Black Anthropocenes or None*, 61.

56. Ibid., xii.

57. Chakrabarty, "The Climate of History," 216.

58. See Serres, *The Incandescent*, 210.

59. Ibid., 213.

60. Ibid., 218. Serres adds this warning to the powerful: "Armed, the strong boast about spreading their peculiar ways to the farthest distances: masters in the morning but dead beat in the afternoon. The weak trust in the Universal, luxuriously varied across space and time" (219).

61. See Jason W. Moore, "The Rise of Cheap Nature," in *Anthropocene or Capitalocene? Nature, History and the Crisis of Capitalism*, edited by Jason W. Moore (Oakland: PM Press, 2016), 78–115 (79).

62. Ibid., 92.

63. The Worldwatch Institute distinguishes between the consumer class, the middle class, and the poor. Its 2004 *State of the World* report estimates that there are 1.7 billion people who belong to the consumer class. Of these, 349 million live in Western Europe (89 percent of the population), 271 million (85 percent of the population) live in the United States and Canada, 240 million in China (19 percent of the population), 122 million in India (12 percent of the population), and 34.2 million in sub-Saharan Africa. See Linda Starke (ed.), *State of the World 2004* (Washington: Worldwatch Institute, 2004), 6–7.

64. Christophe Bonneuil, "The Geological Turn: Narratives of the Anthropocene," in *The Anthropocene and the Global Environmental Crisis*, edited by Clive Hamilton, Christophe Bonneuil and François Gemene (London and New York: Routledge, 2015), 17–31 (29).

65. See Bruno Latour, *Down to Earth: Politics in the New Climatic Regime* (Cambridge: Polity, 2018), 5.

66. Following a question that Dipesh Chakrabarty raises and then refutes in "The Climate of History: Four Theses," 216.

67. See Bonneuil and Fressoz, *The Shock of the Anthropocene*, 71–72.

68. See Franz Mauelshagen, "Bridging the Great Divide: The Anthropocene as a Challenge to the Social Sciences and Humanities," in *Religion in the Anthropocene*, edited by Celia E. Deane-Drummond, Sigurd Bergmann and Markus Vogt (Eugene: Cascade Books, 2017), 87–101 (92).

69. See Amitav Ghosh, *The Great Derangement: Climate Change and the Unthinkable* (Chicago: University of Chicago Press, 2016), 146.

70. See Ibid., 87.

71. See the much-discussed argument of Chakrabarty, "The Climate of History," 221.

72. The Marxist eco-activist Ian Angus therefore argues that treating capitalism and the Anthropocene as identical "can only weaken efforts to get rid of capitalism and mitigate the harm it has caused to the Earth System." See Ian Angus, *Facing the Anthropocene: Fossil Capitalism and the Crisis of the Earth System* (New York: Monthly Review Press, 2016), 232.

73. I therefore do not concur with Eileen Crist that Anthropocene discourse reinforces anthropocentrism and necessarily makes human dominion appear acceptable. See Eileen Crist, "On the Poverty of our Nomenclature," in *Anthropocene or Capitalocene? Nature, History and the Crisis of Capitalism*, edited by Jason W. Moore (Oakland: PM Press, 2016), 14–33. Her position is evident from formulations such as "The Anthropocene discourse clings to the almighty power of the jaded abstraction 'Man' and the promised land his God-posturing might yet deliver him, namely a planet managed for the production of resources and governed for the containment of risks" (23). On this basis, Crist concludes, "such is the poverty of our nomenclature to bow once more before the tedious showcasing of Man. To offer a name which has no added substantive content, no specific empirical or ethical overtones, no higher vision seconded within it—beyond just Anthropos defining a geological epoch" (27).

74. Hamilton, *Defiant Earth*, 58.

75. Ibid., 84.

76. Ibid., 60.

77. Ibid., 77.

78. Ibid., 61. See also the third thesis of Dipesh Chakrabarty in "The Climate of History: Four Theses": "The geological hypothesis regarding the Anthropocene requires us to put global histories of capital in conversation with the species history of humans" (212). He acknowledges that thinking of species as a historical category is dangerous, given historical examples of the political use of biology.

79. Hamilton, *Defiant Earth*, 30.

80. See Naomi Klein, *This Changes Everything: Capitalism vs the Climate* (New York: Simon & Schuster, 2014), 412.

81. Hamilton, *Defiant Earth*, 32, 33.

82. Ibid., 34.

83. Ibid., 58.

84. Ibid., 85.

85. See Frank Biermann et al., "Down to Earth: Contextualizing the Anthropocene," *Global Environmental Change* 39 (2016): 341–50.

86. For Yusoff the Anthropocene is therefore materially constructed as antiblackness, built on the labor of those excluded from its wealth accumulation but who nevertheless have to bodily absorb its waste products as toxicity and confront the storms unleashed. See *A Billion Black Anthropocenes or None*, 82.

87. Following Hamilton, *Defiant Earth*, 93.

88. Dipesh Chakrabarty, "The Politics of Climate Change is more than the Politics of Capitalism," *Theory, Culture & Society* 34:2–3 (2017): 25–37 (28).

89. See Ghosh, *The Great Derangement*, 115.

90. See the formulation by Bonneuil and Fressoz, *The Shock of the Anthropocene*, 229: "To speak of a 'Capitalocene' signals that the Anthropocene did not arise fully armed from the brain of James Watt, the steam engine and coal, but rather from a long historical process of economic exploitation of human beings and the world, going back to the sixteenth century and making industrialization possible."

91. Gabrielle Hecht, "The African Anthropocene," *Aeon Newsletter*, February 6, 2018, https://aeon.co/essays/if-we-talk-about-hurting-our-planet-who-exactly-is-the-we (accessed July 17, 2018).

92. See the argument of Slavoj Žižek in *Living in the End Times* (London and New York: Verso, 2011), 334.

93. A point also made by Chakrabarty, "The Climate of History," 218.

94. See especially Conradie, *Redeeming Sin?* partly drawing insights from Reinhold Niebuhr.

95. See Roy Scranton, *We're Doomed. Now What? Essays on War and Climate Change* (New York: Soho Press, 2018), 8.

96. I am indebted to Frederick Simmons for this insight that he terms the "Anthropocene Paradox." See his unpublished paper "The Anthropocene Paradox and what it Means for Humanism," delivered at the American Academy of Religion, San Diego, November 2019.

97. See Erle C. Ellis, "Neither Good Nor Bad," *New York Times*, May 23, 211, https://www.nytimes.com/roomfordebate/2011/05/19/the-age-of-anthropocene-should-we-worry/neither-good-nor-bad (accessed April 12, 2019).

98. See David Wood, *Deep Time, Dark Times: On Being Geologically Human* (New York: Fordham University Press, 2019), 58.

99. See William J. Ripple et al., "World Scientists' Warning of a Climate Emergency," *BioScience* biz088 (November 5, 2019), https://doi.org/10.1093/biosci/biz088 (italics added).

100. See Dale Jamieson, *Reason in a Dark Time: Why the Struggle against Climate Change Failed and What it Means for Our Future* (New York: Oxford University Press, 2017), 86.

101. This observation is made by Bruno Latour, *Facing Gaia*, 43.

102. See Latour, *Facing Gaia*, 74.

103. See Ibid., 117.

104. See Ibid., 118, with reference to Gilles Deleuze and Felix Guattari, "10,000 B.C.: The Geology of Morals (Who Does the Earth Think It Is)," in *A Thousand Plateaus: Capitalism and Schizophrenia* (Minneapolis: University of Minnesota Press, 1987).

105. See the provocative comment by Latour, *Facing Gaia*, 149–50: "Whereas in the seventeenth century it was necessary, according to Hobbes, that matter be declared inanimate so that order could be re-established, at the beginning of the twenty-first the Earth's retroaction in response to our behaviors suffices to disrupt order thoroughly."

106. Hamilton, *Defiant Earth*, 9.

107. Ibid., 149.

108. See, especially, Grinspoon, *Earth in Human Hands*, xvii, and his discussion of the debate, 241–44.

109. Erle C. Ellis, "The Planet of No Return: Human Resilience on an Artificial Earth," *Breakthrough Journal* 2 (2011): 39–44 (43). The aggressive tone of possessiveness coupled with stewardship adopted here is quite remarkable.

110. Erle C. Ellis, *Anthropocene: A Very Short Introduction* (Oxford: Oxford University Press, 2018), 144.

111. Ibid., 157. Ellis acknowledges that he is accused by some to have coined the term "good Anthropocene but shifts the "culpability" to Andrew Revkin who also coined the term Anthro(po)cene (156). Ellis draws here on the work of Steven Pinker, *The Better Angels of Our Nature: Why Violence Has Declined* (Penguin, 2011). If it is indeed true that, on average, things have never better for the species in human history, especially in terms of food security and health, how is this to be reconciled with the increasing gap between the wealthy and the poor, both within countries and between countries? Perhaps the poorest quintile of the human population should serve as the test case and not the average? Such questions around economic inequality are obviously crucial but also beyond my scholarly competence to address. It is noteworthy that the interaction between poverty and the environment is recognized by Will Steffen and colleagues in the landmark report on *Global Change and the Earth System* to be of local, regional and global significance (128).

112. Ellis, *Anthropocene*, 145.

113. Ibid., 157.

114. See Hamilton, "The New Environmentalism will Lead Us to Disaster."

115. See Elizabeth Kolbert, *The Sixth Extinction: An Unnatural History* (New York: Picador, 2014).

116. Hamilton, Bonneuil and Gemene (eds), *The Anthropocene and the Global Environmental Crisis*, 9.

117. Simon Dalby, "Framing the Anthropocene: The Good, the Bad, and the Ugly," *The Anthropocene Review* 3.1 (2016): 33–51 (48).

118. See Žižek, *Living in the End Times*, 358. He adds: "Does the true not lie in that attitude of total blindness to the achievements of modern secular permissive societies, from women's rights to religious tolerance and the fight against racism?"

119. See Neiman, *Evil in Modern Thought*, 8.

120. See Wood, *Deep Time, Dark Times*, 98, in conversation with Timothy Morton.

121. See also Serres, *The Natural Contract*, 57.

122. See Ibid., 59: "Knowledge asserts the right to contest law. It has so contested it that it has killed it. The death of God amounts to that of the legislator."

123. See *An Ecomodernist Manifesto*, signed by eighteen coauthors, http://www.ecomodernism.org/ (accessed January 20, 2019).

124. Among many others, see the contributions by investigative journalists such as Naomi Klein, Elizabeth Kolbert, Mark Lynas, Bill McKibben, and Roy Scranton.

125. See my proposal for multidisciplinary collaboration to engage in such social diagnostics, the argument that Christian theology may contribute to such diagnostics, and the proposal that the Christian notion of "sin" may be understood, at least from the outside, as a diagnostic of what is wrong in the world. See Conradie, *Redeeming Sin?* It is interesting to note that Johan Galtung also makes use of a diagnosis–prognosis–therapy triangle to suggest that peace studies and health studies are analogous. See Johan Galtung, *Peace by Peaceful Means: Peace and Conflict, Development and Civilization* (Oslo: International Peace Research Institute, 1996), 1, 24–27. See also Slavoj Žižek, *Trouble in Paradise: From the End of History to the End of Capitalism* (Brooklyn: Melville, 2014), 7. He makes a finer distinction between *diagnosis* (of the capitalist system), *cardiognosis* (ideologies that make us accept that system), *prognosis* (what awaits us beyond the end of capitalism), and *epignosis* (knowledge that engage us toward the next phase of the emancipatory struggle).

126. See Stephen Gardiner, *A Perfect Moral Storm: The Ethical Tragedy of Climate Change* (Oxford: Oxford University Press, 2011), 244.

127. Latour, *Facing Gaia*, 21.

128. See Mary Midgley, *Beast and Man: The Roots of Human Nature* (Ithaca: Cornell University Press, 1978), xxxii. See also the comment by Susan Neiman: "I came to philosophy to study matters of life and death, and was taught that professionalization required forgetting them." See *Evil in Modern Thought*, 13.

129. Peter Sloterdijk as quoted in Roy Scranton, *Learning to Die in the Anthropocene: Reflections on the End of a Civilization* (San Francisco: City Light Books, 2015), 27.

130. Jeffrey Sachs promotes what he calls a "clinical economics," one that is alert to the "symptoms" of the "patient" and that "diagnoses" development failures. See Sachs, *Commonwealth*, for example, 15, 223–26; also his *The End of Poverty: Economic Possibilities for Our Times* (New York: Penguin, 2005).

131. Hamilton, *Earthmasters*, 123.

132. See Wood, *Deep Time, Dark Times*, 29.

133. See Žižek *Living in the End Times*, 311.

134. See Hamilton, *Earthmasters*, x.

135. Scranton, *Learning to Die in the Anthropocene*, 21, 23. Scranton adopts a Stoic/Buddhist view in this regard, that implies learning to let go of self, future, certainty, attachment, the pursuit of pleasure, permanence, and stability, to let go of salvation and of hope (92). For Christians, dying means to yield one's life into the hands of a God of hope and of love, or something like that.

136. Jamieson, *Reason in a Dark Time*, 236.

137. See Sachs, *Commonwealth*, 313.

138. See Karl Barth, *Church Dogmatics IV.1* (Edinburgh: T&T Clark, 1956), 138–42, also quoted in Neil Messer, "Sin and Salvation," in *Systematic Theology and Climate Change: Ecumenical Perspectives*, edited by Michael Northcott and Peter M. Scott (New York: Routledge, 2014), 124–40 (126).

Chapter 2

A Theological Contribution?

THEOLOGICAL RESERVATIONS

In this contribution, I will explore the question of what Christian theology could conceivably contribute to such a multidisciplinary engagement with an appropriate diagnosis of what is wrong in the world. In prophetic theology, this task is typically referred to as "reading the signs of the time." How should the significance of this moment in history, this kairos, be understood? For prophetic discourse, this question is tied up not only with a diagnosis of what is wrong with the world but also with a critique of constellations of political, economic, and religious power (speaking truth to power), with the articulation of an alternative vision for the future (a different world is possible), with discerning the countermovements of God's Spirit and therefore the will of God in this particular moment.[1]

The (penultimate) prophetic task is to discern what God is up to, to find the finger of God in history. This is treacherous terrain, as is well recognized by the South African theologian Jaap Durand in an essay on "the finger of God" in history, with reference to apartheid theology.[2] In hindsight, attempts to discern God's hand in history appear to offer a theological legitimation of some or other constellation of political, economic, and religious power. I will return to this task in the postscript.

In the Anthropocene this way of phrasing the question is indeed ominous: Given the possibility of extinction, where is God heading toward with the human species and with God's creation? At the same time, a theological contribution to social diagnostics cannot focus merely on what is wrong with the world, at least not if it lives from the message of salvation that lies at the very core of the Christian tradition. Indeed, it is only possible to see what has gone wrong in the light of the Light of the world. For Christians, sins that are known are also known to be forgiven.[3]

RESERVATIONS ABOUT A THEOLOGICAL CONTRIBUTION

Besides theological reservations as to whether this task of diagnosing what is wrong in the Anthropocene is worth pursuing, several other suspicions as to whether Christian theology can make any meaningful contribution to such a multidisciplinary engagement with diagnostics in the Anthropocene need to be mentioned:

First, in a secularized academy—in the Western world, if not necessarily in Africa—theological studies have been subjected to marginalization and indeed self-marginalization. Given the need for specialization in a multitude of disciplines, the room for any particular contribution from philosophy or theology seems to be limited. In my view, theological studies may be regarded, at least from the outside if not from the inside, as a form of philosophy addressing a particular set of (ultimate and penultimate) questions on the basis of particular texts and traditions.[4] The question as to whether or not the suspicion, namely, that Christian theology has little, if anything, to contribute that the other humanities cannot, is indeed valid, cannot be resolved here. I suppose that the proof of the pudding here lies in the eating, that is, whether this text can demonstrate the value of such a contribution.

Second, it is debatable whether any generalized diagnosis, even if possible, would be helpful. One may speak of a global pandemic (the "Spanish" flu perhaps, Covid-19 certainly) but the notion of a global disease may well be meaningless. Moreover, as many would insist amid global economic and other inequalities, the problem can hardly be humans in general. The root causes of the economic order that has led to the Anthropocene may be found in the history of imperialism and genocide, slavery, colonialism, and exploitation. The *anthropos* of the Anthropocene is white, male, middle class and of recent European descent (if also from more distant African descent). Of course such a diagnosis also remains highly generalized but clearly the global divides between North and South, "developed" and "developing," industrialized and industrializing, should be factored in. This is still rather generalized, often not taking local complexities into account. There is a counterargument here, namely, that human impact in the Anthropocene is no longer on local ecosystems or even bioregions but on Earth systems (the complex interplay between various planetary forces). If so, the diagnosis needs to address Earth systems.[5] Can theology be of use here? Perhaps not, but it is worth reminding one another about the traditional role of myths that are able to express such complexity in a narrative form so that children, even adults can understand what is at stake.

Third, the specific diagnosis typically offered from within the Christian tradition, namely, that human sin lies at the root of the problem, is highly contested—to the point of being subjected to public ridicule. I have analyzed the obstacles that thwart any attempt to retrieve Christian discourse on sin in the public sphere elsewhere under especially five rubrics: (a) cultural resistance, (b) conceptual confusion on the very notion of sin (i.e., the kind of category that it may be), (c) the plausibility of the fall of humanity in the light of evolutionary history, (d) the alleged universality of sin given the need for justice (and hence a distinction between sinners and those sinned against), and (e) a series of classic theological problems regarding sin that remain unresolved on the inside and thus inhibit any public retrieval.[6]

One needs to admit that the track record of diagnosing underlying problems in terms of human sin has been quite disastrous to say the least. Byron Williston refers to "peccatogenetic explanations," namely, explanation of events such as the little ice age or the Lisbon earthquake as originating in human sin.[7] In response, numerous citizens have been accused of, and sentenced to death for, witchcraft, sorcery, and the like, without coming to terms with the root causes of the problem. This critique of invoking the category of sin may be vivid, but the question is whether that also holds for anthropogenic climate change. If humans are to be held morally responsible, can the language of guilt, debt, shame—and therefore sin—be avoided?

Fourth, there is the added problem that such a diagnosis in terms of sin may be presupposed before the patient is even examined. This is a perennial danger for Christian theology, but it is not unique to the discipline. Arguably, the same applies to the paradigms within which social theorists, psychologists, political analysts, and economists operate. Instead, one may suggest that the expertise developed within each of these disciplines allows for a highly differentiated naming of possible diseases on the basis of a set of conceptual tools. Doctors may encounter diseases that their handbooks do not make provision for, but that does not imply that such handbooks are not handy. In the case of Christian theology there is a long-standing vocabulary that has been developed in pastoral and in prophetic theology that may be utilized whenever a new "patient" is examined, that is, whenever the need arises to discern the "signs of the time" anew. The much maligned seven deadly sins of medieval Christianity may serve as an example of a pedagogical tool to recognize when things do go wrong, at least at a personal level.

Fifth, and this is perhaps where the real difficulty lies, for an adequate diagnosis of the Anthropocene the question is what is really new. What has changed that prompted the naming of the Anthropocene? Hunger, poverty, inequality, and injustice have formed part and parcel of human history. Likewise, the ecological impact of new technologies is nothing new. What is it, then, that has taken things over the edge to warrant a new diagnosis? The medical analogy remains instructive: would the tests reveal that the growth is malignant or not? Is the disease indeed "terminal"? This is not only a complex question but framing it in this way also suggests an apocalyptic urgency with which the verdict is awaited, as is the case with the verdict following medical tests. The problem with apocalyptic discourse is that it all too often invokes fear instead of hope—the latter being the main orientation of apocalyptic literature in the biblical witnesses amid political oppression and surveillance from security forces. There is no shortage of apocalyptic symbols in the Anthropocene, as is obvious from contemporary literature, films, graphic art, and music. Ironically, such frightening images are mass produced for entertainment. The bitter paradox is that the use of apocalyptic genres seems more at home in religious fundamentalism than in prophetic discourse, and typically induces paralyzing and escapist fear, together with anxiety, morbidity, fatalism, and annihilating nihilism. Here, too, religion (especially right-wing evangelicalism) remains part of the problem.

Finally, there is the question whether theology (or more specifically the message of the gospel that it reflects on) can offer any hope to address the underlying problems as diagnosed. The hope of any diagnosis is of course that to identify and specify the root causes of the disease (and not just the symptoms) will enable one to suggest, test, and where need be modify an appropriate remedy or therapy, whether preventive or curative. The diagnostic exercise may sound rather negative but "going to the doctor" is the first step to overcome the disease. It is of course possible that the diagnosis may be devastating, that the disease is indeed terminal. Such a verdict is often found in secular literature too.

However, the deeper suspicion may be that even if an accurate diagnosis may become available, the Christian gospel does not offer much of a solution. Its message of salvation has been tried and tested and found wanting. It may still have some functional value to provide moral inspiration to adherents where need be, but what is needed in the Anthropocene is not a reiteration of religious remedies that have not resolved so many other social problems. One may retort that the same applies to technological, economic, and political remedies but this does not alleviate the burden of proof. What, then, can Christian theology contribute to a diagnosis of what is wrong with the world amid the Anthropocene?

TWO TOOLKITS

In this contribution, I will engage with the question of what is wrong with the world as explored in secular discourse on the Anthropocene. I will assume that such a question can only be addressed in the public sphere through multidisciplinary dialogue in which the academy is only one stakeholder. I will explore the question as to what Christian theology can contribute to this task. I acknowledge that the most significant contribution may well be to merely participate in the debate, to make one's bodily presence felt, to listen carefully to the contributions of others, to help insist that marginalized voices are heard. Yet, in order to participate in such a debate on diagnostics, one needs to be prepared to answer the question from other participants: What do you have to contribute? To refrain from addressing such a question may be insincere and vulnerable to accusations of tacitly seeking to play the role of facilitator, scribe, chairperson, or even legislator. Moreover, it would be irresponsible to fail to make a contribution that is arguably much needed and where no other disciplines may be able to make that very contribution.

I will bring to the table especially two toolkits, both derived from theological discourse:

CHRISTIAN VOCABULARY ON SIN

The first toolkit comes from the long-standing tradition of Christian sin-talk. In this tradition, a fairly sophisticated (but nevertheless much maligned) vocabulary has been developed to diagnose the underlying problem by naming several manifestations of human sin. Each of these has been picked up in environmental discourse with secular equivalents. Let me simply list these here in order to return to such tools whenever useful:[8]

- For some, anthropocentrism forms the root cause of ecological destruction in the Anthropocene (see sin as pride, hubris, "playing God").
- For others, consumerist greed, especially but not only among the global consumer class, is the problem (see sin as greed, lust, concupiscence, selfishness, mimetic desire, misdirected longings).[9]
- Yet others would argue that our collective failure to accept (political) responsibility is to be blamed for allowing things to go beyond control, for tolerating poverty, unemployment, and inequality, for culpable negligence (see sin as "missing the target," i.e., *hamartia*, sloth, false humility,[10] neglect, apathy, also blame-shifting, scapegoating).
- Some would say that it is the modernist disconnection from nature,[11] the disenchantment of the world, alienation from the sacred soil, from the Earth as our home, the divide between nature and culture, indifference to the needs of others, that has allowed ecological destruction because nonhuman nature "out there" is deemed to be "ripe for exploitation" (see sin as privation of the good).
- Many would insist that domination in the name of differences of gender, race, class, language, and culture constitutes the root of the problem (see

sin as violence, as the violation of human dignity, subjugation, dispossession, extraction). The multiple forms of domination through the extractivist commodification of natural "resources" and the exploitative division of labor are explored in many available critiques of capitalism, at best in the context of ecofeminism.

- One may also speak of moral corruption that spreads like a virus to undermine the validity of the rules maintained in society, that perverts values, favors seductive alternatives, creates a mood of lawlessness, and enfeebles the will to abide by such rules (see sin as distortion, *corruptio totalis*, trapping people in a mess from which they cannot extricate themselves). This description of sin builds on the privation of the good and specifies its impact.
- Then there is also the obvious role of pollution, albeit that the problem is not only the biophysical impact of such pollution but the defilement of hearts, minds, and habits that allow such pollution to proliferate, to be condoned, to be not even noticed as such (sin as blemish, dirt, toxic, impurity, sacrilege, desecration, stain).[12] Again this builds on sin as the privation of the good.
- Finally, one may say that a lack of wisdom pervades contemporary discourse in the Anthropocene, despite the presence of so many expert advisors (see sin as folly, falsity, denial, ignorance, deceit,[13] hypocrisy).[14]

The value of this toolkit is partly that it does not dwell on the surface-level symptoms of the underlying problem. It also does not focus on the many factors that may contribute to the problem. In the Anthropocene these could be related to human abilities, for example, for intelligence and self-consciousness, language and communication, or creativity and technology. These can evidently be used for doing evil *and* good. This may help to overcome endless debates on whether the side effects of technological progress can be addressed by more technology. Or whether we can think our way out of the problem if the problem is the way we think.[15] All these capacities are clearly needed but overreliance on any one would underestimate the depths of the predicament.

This toolkit seeks to capture where things can and do go wrong, albeit that this is often focused at the individual level, thus underplaying the communal, the collective, the institutional, and especially the structural dimensions of evil. Individual moral wrongdoing is perhaps the common sense understanding of sin, but the limitations of such a view are obvious from the following list of conflicting interpretations of sin:[16]

- Sin describes individual acts of human wrongdoing: what individuals do to others, to themselves, and to their immediate environment by not doing what is right, not abiding by rules (anomie).

- Sin is not just a matter of doing, but also of thinking and saying. It is about dispositions, attitudes, attachments which provide the breeding ground for wrongdoing. Sin affects the mind, the imagination, the sense of longing, desires, feelings, the will, the conscience, and therefore also the senses, each part of the human body.[17]
- Sin also refers to what is left undone, a failure to accept responsibility, duties not fulfilled.
- Sin describes the character of a person rather than particular things that a person may do. Being a sinner is a more pervasive problem than committing acts of sin. Habitual sin is much harder to stop.
- Sin describes the quality of a (broken) relationship of trust and loyalty, not the dispositions or deeds of any one individual. Selfishness is a manifestation of a broken relationship. If so, one cannot sin on one's own. Sins in the plural are manifestations of broken relationships.
- Such broken relationships are embedded in wider networks of families, clans, institutions, and affiliations. Sin becomes manifest in what one party does to another but is indeed best understood as a relationship going awry.
- Sin is best understood as structural violence; it describes systems of oppression such as patriarchy, slavery, colonialism, apartheid, castes, capitalism, ecological destruction, a situation in which individuals and organizations alike are caught up.
- This may be true, but then the focus should be on the power of ideological distortions: sin is about classism, sexism, racism, elitism, homophobia, xenophobia, anthropocentrism, and so on.
- Sin cannot be reduced to something moral but is primarily something religious. It is indeed about broken relationships but then a broken relationship with God, broken at least from our side.
- Sin is best understood as idolatry. If it is religious in nature, it is not merely about an individual's relationship with God but about unbelief, rebellion, idolatry, putting one's trust in principalities and powers that cannot save us.
- Sin is about heresy, about radically distorting the Christian gospel to serve one's own interests.

What kind of thing is sin, then? The answer may well be that sin is not something that one can identify and describe but rather the privation of the good. This Augustinian position assumes the ontological priority of the good so that evil is a distortion of such good. However, to explore sin may be fraught with many dangers. In Christian reflection on the nature of sin, the same conclusion is reached again and again, namely, that sin cannot be defined but can only be opposed.[18] The attempt to define sin is already part of the problem. It may also underestimate the way in which sin morphs into different manifestations. If it could be captured in a single formula it might have been possible to isolate, contain, and surgically remove sin like a cancerous growth. Instead, sin is seen as elusive and hideous, reemerging in new forms.

Another way of reflecting on these notions of sin is in terms of the classic distinction between sin as guilt and as power.[19] Some would focus on sin as guilt and then attend to things that individuals do. Others would focus on sin as power and then attend to how individuals are influenced by forces beyond their locus of control. We are caught up amid evil forces that are more powerful than what we can cope with and from which we cannot escape. A one-sided emphasis on either is dangerous. To focus on individual actions is to trivialize sin, to look for particular instantiations of sin as if there are other actions, attitudes, dispositions, and thoughts that are not contaminated by sin. To focus on sin as power is only to portray individuals and groups as victims of forces beyond their control for which they therefore ultimately do not need to accept responsibility. This would undermine agency, also the agency of victims.

I do think that it is possible to combine an understanding of sin as power and as guilt in order to avoid such excesses. Perhaps one may start with sin as power (structural and cultural/ideological violence). Accordingly, sin may be understood as pervasive perversity. It is the mess that we find ourselves in. The Anthropocene may serve as another name for that mess, while the "good" Anthropocene" is then a sinister attempt to render the unacceptable acceptable.

One would need to add that this is a mess to which we (humans) and our predecessors have all contributed in one way or another to create or to worsen the mess. However, there is no need to argue that we have contributed equally. Consider the role of those moral agents in positions of political and economic power but also those moral patients (infants, the sick, the senile) on whose behalf others act. The victims of history are seldom completely innocent and may well become the perpetrators of tomorrow. If we are all equally guilty, then we can no longer introduce gradations to establish who is truly guilty or not. Then moral appeals cannot and do not make an appeal.[20]

Second, this is a mess that causes suffering to all of us, including other animals and plants, albeit again not equally so.[21] The suffering of the victims may be obvious, but tyrants and torturers also suffer the consequences of oppression. However, it would be obscene to equate the suffering of the rapist and the rape victim. This is obviously relevant in the context of climate change where those who contribute next to nothing to carbon emissions will suffer disproportionally under its impact.

Third, Augustinians (more than Pelagians) would add that this is a mess from which we cannot escape, at least not by ourselves. Our best efforts at moral renewal, reconstruction, transformation, and social development, remain flawed and often even exacerbate the problem. In short, sin describes the mess that we find ourselves in, to which we all contributed (but not equally so), under which we all suffer (again not equally so) and from we cannot escape.

THE THEOLOGICAL CRITIQUE OF APARTHEID

The second toolkit comes from the South African context where I am based and more specifically from the theological critique against apartheid developed, especially in the 1970s and the 1980s. It was widely recognized that apartheid was far more than merely a political policy or system; it entailed a way of thinking (consciously or subconsciously), an ideology, a form of life, the way of arranging society often to the smallest details of whom you may marry, where you may live, where you may buy, where you may swim, in which queues you should stand, which toilets you may use. This way of thinking was typically supported, motivated, and popularized with reference to biblical and theological motifs.[22] Accordingly, multiple layers of apartheid had to be confronted, which, I suggest, may be correlated with a differentiated notion of what sin entails. Let me describe these only briefly and with reference to the critique of apartheid in the 1980s in order to return to each of these layers in the subsequent discussion:

First, for some, apartheid simply entails wrongdoing through racial discrimination in numerous specific contexts. In each instance, one may speak of injustices being perpetrated, of dignity being violated. Such discrimination may be overt or covert, conscious or subconscious, more or less brutal, with consequences on the shorter or the longer term. Insofar as such discrimination still continues, apartheid is far from over, so that the present tense is used advisedly. Moreover, one may recognize that sin may become habitual, taken for granted, and therefore extended over longer periods of time than is the case with individual acts of wrongdoing. Such "habits of the heart" lead to racial prejudices (and xenophobia) and provide the breeding ground for racial discrimination. It becomes the accepted "way in which we are doing things around here."[23] In the apartheid era, there were many cases where such racial discrimination amounted to both minor and major violations of human rights. The South African Truth and Reconciliation Commission focused (only) on such major violations of human rights, often associated with police brutality, torture, and murder. This left untouched the countless cases where minor violations of human dignity were experienced by ordinary South Africans.

Second, one may add that such practices embedded in cultural patterns can also become institutionalized in laws, policies, and implementation strategies. Apartheid entailed a far-reaching geographic reconstruction of society, touching on issues of marriage and sexuality, housing, education, health, and the political economy. More specifically, it structured the South African economy in such a way that job opportunities, contracts, loans, housing, education, health, and social services were unequally distributed.[24] Such inequality implied that specific sectors of the population ("whites," later to a limited extent also "Indians" and "Colored" people) could be regarded as the beneficiaries of apartheid, while others were systematically disadvantaged. These beneficiaries could easily take their privileged position for granted as inherent to the social order, so that awareness of the injustices embedded in that social order was ideologically repressed. This amounts to structural violence that prompts an ongoing spiral of revolutionary violence and repressive violence (sin as oppression). The legitimate role of the state to curb the spread of "evil" (e.g., terrorism) through law, policing, and the military was rigorously defended in apartheid theology.[25] Even when apartheid was no longer a viable political policy, the ideology of "state security" could prevail. This created a dispensation, a total mess in which oppressor and oppressed became trapped and from which it became increasingly hard to escape (see sin as power).

Third, many have added that sin becomes manifest in pervasive ideologies (racism, capitalism, classism, sexism, white supremacy). Ideologies need to be named precisely because they tend to remain hidden from the consciousness of those who harbor such ideologies. They allow one to make assumptions, to take a particular social order for granted, to do things without realizing what impact this may have on others. They operate as radical distortions of consciousness. Where such ideologies do become self-conscious, one may also speak of sin as disobedience, even as rebellion, refusing to abide by what one knows to be the right thing. While it is best to speak of several contending ideologies in South Africa in the 1980s,[26] several critics exposed the ideological underpinnings of the politics of apartheid in terms of racism, classism, and nationalism, including Christian nationalism.[27]

Fourth, apartheid also functioned as a pseudo-gospel[28] and a quasi-soteriology, that is, as a functional replacement for God's salvation.[29] Most basically, it became manifested in the belief that the only way to protect the "flame" of European "civilization" amid the "sea of barbarism" in Africa is by setting oneself apart and protecting cultural identity through economic, political, and military power. Elsewhere, education, health care, pension funds, and insurance operate as functional and institutionalized replacements for the need to be "saved" from within dire circumstances.

Fifth, a closely related critique was that apartheid may be regarded as a form of idolatry, that is, the willingness to serve other gods, often in the form of syncretistic mixtures between faith in the God of Israel and of Jesus Christ and an array of pagan gods (nationalism, white supremacy, prosperity, and fertility cults, indeed religion itself). In response, drawing especially on the work of Karl Barth and Dietrich Bonhoeffer, the theological critique against apartheid insisted, as captured by John de Gruchy, that "God was not the God of religion, culture and nation, but the God who judges such idolatry, calling both church and nation to repent and pursue God's justice."[30]

Finally, apartheid was also critiqued as a heresy, that is, as the more explicit theological legitimation of an oppressive political system.[31] In response to such heresy, the need for a confession of faith (*status confessionis*) against distortions of that faith was declared. This prompted what was known as a movement toward a confessing church at that time. I will return to this aspect of the critique against apartheid in more detail below.

Together these layers of the critique against apartheid constitute six deadly sins of a different order where each next layer is arguably deadlier than the previous one. My hope is that this toolkit, in particular, may allow for an authentic contribution from the global South to global discourse on the Anthropocene.

SECULAR DISCOURSE ON SIN

In what follows below, I will explore the appropriation of these two toolkits in the context of secular discourse on the Anthropocene. Could this help to develop a more sophisticated and differentiated diagnosis of what has gone wrong with the world? In general, I will focus only on those aspects where the disciplines of philosophy, ethics, religious studies, and Christian theology can contribute insights that other disciplines cannot. In other words, I will not focus on scientific (e.g., geological) or economic discourse on the Anthropocene, even though this obviously has to be taken into account. I will do so by offering a theological analysis and critique of the available secular literature, that is, by using theological tools without offering any detailed exposition of such tools. This strategy avoids repetition albeit that a clinical separation between disciplines is precisely not possible in the Anthropocene.

Literature on the Anthropocene has quickly expanded and can no longer be surveyed. This applies especially to the natural sciences but also to the humanities and social sciences. I adopted a strategy to find appropriate literature that can at best be regarded as snowball sampling, that is, finding one appropriate text that leads one to others. This is imperfect but would suffice for present purposes.

Often the selected disciplines offer a critical interpretation of data, knowledge, and insights emerging from within other disciplines. Nevertheless, my strategy will be to offer a theological reading of *secular* discourse on sin, that is, by showing either the tacit theological assumptions in such literature or to what (limited) extent theological categories can illuminate such literature. In other words, while the tools that will be used are theological by origin, such tools will be used to diagnose the underlying predicament in material that is by and large not explicitly of a theological nature. In the process, the relative adequacy of such tools for public discourse on the Anthropocene will be open to scrutiny.

One possible misunderstanding of the aim of this contribution should be noted. It would be facetious to assume that each and every of the layers in the critique against apartheid mentioned above will indeed be applicable to discourse on the Anthropocene. In this sense the question is how far this analysis of the layers of the problem can be extended. It would be inappropriate to assume that it can be extended all the way, although looking for some evidence would be appropriate. One would find more common ground for a critique of injustice and also of hidden ideologies at work in the Anthropocene than for a critique of idolatry and heresy. At the same time, it is unlikely that "injustices" can capture what is wrong in the Anthropocene. The problem surely lies deeper than that. There should be some middle ground as far as quasi-soteriologies are concerned given activist and geoengineering attempts to "save the planet." On the interplay between economic injustice and ecological degradation, several scholars have observed forms of idolatry that are relevant also within the Anthropocene, but few have suggested that some form of heresy is at stake. A twofold critique of Christian discourse is therefore needed here, namely, to recognize the limits of its own discourse but also to test the plausibility of its own claims for salvation.

NOTES

1. For one South African discussion of this mode of prophetic theology, see Albert Nolan, "Theology in a Prophetic Mode," in *Hammering Swords into Ploughshares: Essays in Honour of Archbishop Mpilo Desmond Tutu*, edited by Buthi Tlhagale and Itumeleng Mosala (Johannesburg: Blackshaws, 1986), 131–40.

2. See Jaap Durand, "God in History: An Unresolved Problem," in *The Meaning of History*, edited by Adrio König and Marie Henri Keane (Pretoria: Unisa, 1980), 171–78.

3. See again Messer, "Sin and Salvation," 126.

4. For a more detailed argument, see Ernst M. Conradie, "Twelve Theses on the Place of Christian Theology in Multi-disciplinary Conversations," *Stellenbosch Theological Journal* 1:1 (2015): 375–86.

5. See the comment by Steffen et al., "Trajectories of the Earth System," 1: "While recognizing that different societies around the world have contributed differently and unequally to pressures on the Earth System and will have varied capabilities to alter future trajectories, the sum total of human impacts on the system needs to be taken into account for analyzing future trajectories of the Earth System."

6. See Conradie, *Redeeming Sin?* Chapter 4.

7. Byron Williston, *The Anthropocene Project: Virtue in an Age of Climate Change* (Oxford: Oxford University Press, 2015), 5.

8. I have analyzed some of these environmental manifestations of sin in more detail in "Penultimate Perspectives on the Root Causes of Environmental Destruction in Africa"; also *Redeeming Sin?* Chapter 2.

9. I find the following comment by the American Geophysical Union based on focus group discussions (conducted in 1998) on attitudes toward global warming rather chilling: "Many respondents in our focus groups were convinced that the underlying cause of environmental problems (such as pollution and toxic waste) is a pervasive climate of rampant selfishness and greed, and since they see this moral deterioration to be irreversible, they feel that environmental problems are unsolvable." Quoted in Klein, *This Changes Everything*, 60. Klein bemoans the neoliberal success in persuading us that we are selfish, greedy, self-gratifying machines, "not just incapable of self-preservation but fundamentally *not worth saving*" (62). She counters this with a Pelagian insistence that we are a "mess of contradictions" where self-gratification coexists with deep compassion and solidarity. The Augustinian view is more radical though, namely, that even though we are hopelessly trapped in selfish greed we are not only worth saving but worth dying for, as the Gospel according to John 3:16 suggests.

10. This is the argument of Mark Lynas in his controversial book *The God Species: Saving the Planet in the Age of Humans* (Washington: National Geographic Society, 2011). Counter to the "green creed" that "playing God" is dangerous, Lynas argues that false humility is a more urgent danger than hubris. Playing God in the sense of being intelligent designers is essential in order to avoid irreparable damage to planetary boundaries (10). We are able to dominate and conquer but also to nurture and protect (13) our (?) ecological capital assets like rain forests that are "worth more alive than dead" (12).

11. See Clive Hamilton, *Requiem for a Species: Why We Resist the Truth about Climate Change* (London and New York: Routledge, 2015), Chapter 5, 134–58.

12. See, for example, Michel Serres's analysis of pollution as the desecration of places of worship by some excrement. He traces the extension of such defilement (and appropriation) to trash, sewage farms, waste dumps, industrial waste dumps and, one may add, the oceans and the atmosphere. See Michel Serres, *Malfeasance: Appropriation through Pollution* (Stanford: Stanford University Press, 2011), 34.

13. In addition to climate denialism that may have psychological roots, there is also deliberate deceit. See especially Naomi Oreskes and Erik M. Conway, *Merchants of Doubt* (London: Bloomsbury, 2010).

14. For one example in discourse on the Anthropocene, see Grinspoon, *Earth in Human Hands*, 277. He calls for a terra sapiens (wise earth), in the hope for a Sapiezoic era, one "when cognitive processes become a stable part of a planet's functioning," and "when life realizes what it is doing and incorporates that knowledge into the operation of its planet." Calvinist discourse on pervasive depravity would cast some doubt on such an ability to learn from mistakes. What would guarantee such an ability? Grinspoon admits that his views arise from his "faith in the basic goodness of

life, intelligence and consciousness" and that "this may be the best retort against the logical possibility of evil" (350).

15. See Hamilton, *Earthmasters*, 199.

16. For a more detailed discussion, see Conradie, *Redeeming Sin?* Chapter 4, also "To Cover the Many Sins of Galamsey Mining."

17. See Herman Bavinck, *Gereformeerde Ethiek* (edited by Dirk van Keulen) (Utrecht: KokBoekencentrum, 2019), 83, 91. See also Bavinck's superb discussion on the classification of various kinds of sin, in terms of making nature, others, and God serviceable to oneself, either one's body or one's spirit (104).

18. See the comment by Dietrich Bonhoeffer on the related question of the origin of evil: "The question why there is evil is not a theological question, for it presupposes that it is possible to go back behind the existence that is laid upon us as sinners. If we could answer the question why, then *we* would not be sinners. We would blame something else. So the question 'why' can never be answered except by the statement 'that' which burdens humankind so completely. The theological question is not a question about the origin of evil but one about the actual overcoming of evil on the cross; it seeks the real forgiveness of guilt and the reconciliation of the fallen world." See Dietrich Bonhoeffer, *Creation and Fall: A Theological Exposition of Genesis 1–3*, edited by John W. de Gruchy (Minneapolis: Fortress Press, 1997), 120.

19. These few paragraphs draw from my essay Ernst M. Conradie, "The Emergence of Human Sin," in *T&T Clark Handbook on Christian Theology and Climate Change*, edited by Ernst M. Conradie and Hilda P. Koster (London: T&T Clark, 2019), 384–94. This essay was written simultaneously with working on this monograph, with obvious overlaps.

20. See René ten Bos, *Dwalen in het Antropoceen* (Amsterdam: Boom, 2017), 70.

21. Dipesh Chakrabarty makes this point forcefully: "Left unmitigated, climate change affects us all, rich and poor. They are not affected in the same way, but they are all affected. A runaway global warming leading to a Great Extinction event will not serve the rich very well. A massive collapse of human population caused by climate dislocation—were it to happen—would no doubt hurt the poor much more than the rich. But would it not also rob global capitalism of its reserve army of 'cheap' labor on which it has so far depended?" See Chakrabarty, "The Politics of Climate Change," 30.

22. See Dirk J. Smit, "What Does *Status Confessionis* Mean?" in *A Moment of Truth: The Confession of the Dutch Reformed Mission Church*, edited by G. Daniel Cloete and Dirk J. Smit (Grand Rapids: Eerdmans, 1984), 7–32 (30).

23. I borrow this colloquial phrase from conversations with George Hunsberger who uses it with reference to discourse on "the gospel and *our* culture," that is, the way in which the gospel challenges the home culture of cross-cultural missionaries.

24. On the legacy of such inequality, see especially Sampie Terreblanche, *A History of Inequality in South Africa 1652–2002* (Scottsville: UKZN Press, 2002).

25. Johann Kinghorn argues that apartheid theology was actually one of the first forms of public theology to operate with a structural understanding of sin. It saw the diversity of ethnic groups as God-given and any tendency toward integration (symbolized by the tower of Babylon) as in danger of imperialism. This required a

political dispensation where distinct nations (based on ethnicity) can coexist by being held apart. Apartheid theology defended what would later become known as "state theology" on the basis of the classic reformed understanding of the role of the state to contain evil sufficiently so that the message of salvation can have an impact. See Johann Kinghorn (ed.), *Die NG Kerk en Apartheid* (Braamfontein: Macmillan, 1986), 180–84.

26. See especially James Leatt, Theo Kneifel and Klaus Nürnberger (eds), *Contending Ideologies in South Africa* (Cape Town: David Philip/Grand Rapids: Eerdmans, 1986); also Dirk J. Smit, "The Symbol of Reconciliation and Ideological Conflict in South Africa," in *Reconciliation and Reconstruction*, edited by Willem S. Vorster (Pretoria: Unisa, 1986), 79–112.

27. See already Allan A. Boesak, *Afscheid van de Onschuld: Een Sociaal-ethische Studie over Zwarte Theologie en Zwarte Macht* (Kampen: JH Kok, 1977), 87–106.

28. This was already expressed in a document entitled "A Message to the People of South Africa," released by the South African Council of Churches in 1968. One paragraph reads: "There are alarming signs that this doctrine of separation has become, for many, a false faith, a novel Gospel which offers happiness and peace for the community and for the individual. It holds out to men [*sic*] a security built not on Christ but on the theory of separation and preservation of their racial identity. It presents separate development of our race-groups as a way for the people of South Africa to save themselves. Such a claim inevitably conflicts with the Christian Gospel, which offers salvation, bot social and individual, through faith in Christ alone." See the appendix in John W. de Gruchy and Charles Villa-Vicencio (eds), *Apartheid is a Heresy* (Cape Town: David Philip, 1983), 155.

29. For a discussion, see Murray H. Coetzee and Ernst M. Conradie, "Apartheid as Quasi-soteriology: The Remaining Lure and Threat," *Journal of Theology for Southern Africa* 138 (2010): 112–23. This article draws on an extensive literature from the period of struggle against apartheid, including David J. Bosch, "The Roots and Fruits of Afrikaner Civil Religion," in *New Faces of Africa: Essays in Honour of Ben (Barend Jacobus) Marais*, edited by Jan W. Hofmeyr and Willem S. Nicol (Pretoria: University of South Africa, 1984), 14–35; and Jaap Durand, "A Confession – Was it Really Necessary?" in *A Moment of Truth: The Confession of the Dutch Reformed Mission Church*, edited by G. Daniel Cloete and Dirk J. Smit (Grand Rapids: Eerdmans, 1984), 33–41.

30. John W. de Gruchy, *The End is Not Yet: Standing Firm in Apocalyptic Times* (Minneapolis: Fortress, 2017), 62.

31. See, especially, the volume by De Gruchy and Villa-Vicencio, *Apartheid is a Heresy*.

Chapter 3

Unsustainable Habits of the Heart in the Anthropocene?

I=PAT?

One widely used formula (I=PAT), first introduced by Paul Ehrlich in the 1970s,[1] suggests that environmental impact (I) is a function of population (P), patterns of production of consumption (Affluence=A), and the force unleashed by the technologies employed (T). While strictly speaking this formula expresses an identity that holds by definition, it does identify three factors that also lie behind particular surface-level symptoms in the Anthropocene such as global warming, the extinction of species, and ocean acidification (global warming's "equally evil twin"[2]). It may, therefore, offer a point of entry to see what has gone wrong in the world.

The formula correctly indicates that growth in any of these factors would aggravate environmental impact. Admittedly, it hides differences among humans and may lead to dual accusations insofar as some would want to criticize birth rates in the global South while others emphasize the ruinous consumption patterns in affluent societies. In particular, an emphasis on the population variable can have the effect of blaming the victims for consequences such as tropical deforestation and famine-malnutrition.[3] An emphasis on consumption patterns seems to suggest that the affluent must sacrifice their lifestyles and that the global poor, the potential "new consumers" in the so-called developing world must give up their middle-class dreams and aspirations.[4] If neither population nor affluence can be touched as being politically too risky, then the emphasis has to shift to technological innovation. The question is therefore what the relative weight of each of the three factors may be, but this is rather hard to establish. Andreas Malm and Alf Hornborg warn that there is no direct correlation between population and CO_2 emissions:

> Granted, there is a correlation between human population and CO_2 emissions, but the latter increased by a factor of 654.8 between 1820 and 2010 . . . , while the former "only" did so by a factor of 6.6 . . . , indicating that another, far more powerful engine must have driven the fires. For recent decades, the correlation has been revealed as outright negative . . . in the quarter-century between 1980 and 2005 . . . numbers tended to rise fastest where emissions grew slowest, and vice versa.[5]

The demographics of population growth, the thread mill of production, the increasing levels of consumption, especially among the consumer class, and the rapid spread of technological innovation are widely discussed in discourse on the Anthropocene. There is no need to repeat the argument here. In short, by 1945, the great majority of the Earth's human population lived outside of the materials-intensive, high-energy, consumer-driven economy that now dominates the globe. The story of the seven decades since then is one of incorporating ever-larger parts of the world into this economy, even though much of Africa remains excluded and even though many remain on the economic periphery, also in highly industrialized, so-called developed countries.[6] In whatever way one approach this, it should be clear that the resource usage of the most affluent one billion people cannot be copied by all others in a sustainable way.

What difference does the Anthropocene make in this regard? One may suggest that such patterns were always in danger of becoming unsustainable and that the recognition of the Anthropocene is an indication that a tipping point has been reached and breached. This is widely recognized as the "Great Acceleration" that occurred since the 1950s in terms of almost anything that can be measured: population growth, gross domestic product (GDP) growth, carbon emissions, fertilizer production, automobiles, transportation, tourism, ocean acidification, and so on. It focuses attention to carbon, nitrogen, and sulfur cycles that have clearly been impacted by humans collectively. The impact of the Anthropocene will last much longer than the Great Acceleration that will by necessity come to an end; it will last at least as long as the higher levels of carbon in the atmosphere and oceans remain evident, measured in centuries and millennia.[7]

Such acceleration cannot last simply because of planetary limits that, for example, make a further the quadrupling of water withdrawals, a doubling of urban population, or a quintupling of marine fish catch impossible. As the major report on *Global Change and the Earth System* observes, "Exponential changes of the kind currently being experienced cannot go on indefinitely.

Such changes must, in principle, lead to one of three outcomes. The first is stabilization at a new state of the system with its own characteristic patterns of variability. The second is relaxation of the system back to its previous state at a manageable rate, and the third is a more rapid or catastrophic change of the system to its previous state or to a different state."[8] It adds that despite recent advances, it is not possible at this stage to make any projection of how global change will progress. What the report does say is nevertheless disturbing, namely, that human impacts on the Earth system,

> are approaching or exceeding in magnitude some of the great forces of nature; operate on much faster time scales than rates of natural variability, often by an order of magnitude or more; and taken together in terms of extent, magnitude, rate and simultaneity, have produced a no-analogue state in the dynamics and functioning of the Earth System.[9]

Such a notion of planetary limits seems statistically self-evident but requires further unpacking, if only because of the controversies over "limits to growth" that erupted ever since the publication of the report to the Club of Rome in 1972.[10] What limits are at stake in this regard? Limits to population growth? Economic growth?[11] Technological progress? Human ingenuity? More specifically: What is meant by sustainability here and when do human habits really become "unsustainable"? This is not an easy question to answer. In a way, the question regarding discourse on sustainability is what the question really is. This is important in order to pinpoint as far as possible exactly where and why practices become unsustainable. Consider the following significant shifts in the debate that have taken place over the past few decades.[12]

SHIFTS IN DISCOURSE ON SUSTAINABILITY

At first, the question, raised within the context of the original *Limits to Growth* report (1972), was whether or not the use of nonrenewable resources can be sustained (although the term sustainability did not appear in this report yet), given human population growth and an increasing consumption of fossil fuels. This is a nonsensical question since the use of *nonrenewable resources* can, by definition, not be sustained. The more appropriate question is, therefore, how long the use of such resources (including for all practical purposes coal and oil) can be sustained (an important question for policy making). This raises questions around efficiency and unnecessary waste. A related question for engineers is of a technological nature: how can the use of such resources be extended and optimized? Can suitable replacements (e.g., in terms of energy sources) be found or produced?

The focus partially shifted away from such questions with the Brundtland report entitled *Our Common Future* (1987), namely, in order to address the sustainable use of *renewable resources* such as water, fisheries, forests, and especially soil. This suggests that the limits to the sustainable use of renewable resources are more crucial than limits to the availability of nonrenewable resources.[13] In the famous definition of the Brundtland report, sustainability was understood as the use of resources to attend to current needs without compromising the ability of future generations to attend to their own needs.[14] The main concerns here are over the pollution of such resources but also over the ability of ecosystems to replenish themselves, for example, regarding the fertility of the soil. The question is therefore whether current patterns of production and consumption are compatible with the carrying capacity of land, measured by ecological footprint and in terms of biomass.[15] Since biomass can be stored (e.g., in trees or fossil fuels), it is possible to utilize more biomass for human consumption than is produced in a given year.

Another shift, especially since the Rio Earth Summit (1992), was related to concerns over the human victims of the unsustainable use of renewable resources, raising issues over environmental justice and environmental racism. This link between social justice and environmental sustainability was preempted by the World Council of Church at both its Nairobi (1975) and Vancouver (1983) assemblies. One may say that the (rather utilitarian) question is whether the benefits of economic production and consumption for some outweigh the costs for others and whether this pattern can be continued (sustained!) without such social instability that the viability of economic systems is undermined. In the Western Cape, this may be captured in terms of access to sanitation (flush toilets) amid chronic water shortages.[16] In order to counter such a utilitarian calculus, the emphasis on ecojustice is often complemented by the notions of sustainable livelihoods and sustainable communities as an alternative to the sustainable development given its (unsustainable) assumption of sustained economic growth.[17] I will return to this assumption in the next chapter.

An even more significant shift took place with the annual meetings of the Conference of the Parties in the context of the UN Convention Framework on Climate Change, namely, to focus on the absorption capacity of the biosphere. Sustainability is here defined in terms of the ability of the biosphere to absorb the waste products of an industrialized economy, most notably carbon dioxide, but also toxic waste and especially nuclear waste. Everything can be recycled, but this requires time so that the production of waste (i.e., a high concentration of particular chemicals) can exceed the biosphere's absorption capacity. This capacity is determined by the ground rules of chemistry (and not economics!). In the case of greenhouse gases, the absorption takes place through photosynthesis, the oceans, and the atmosphere, leading to flourishing plant life but also to the acidification of the oceans and to global warming. The question thus becomes: how long can excessive waste be produced without jeopardizing the system as a whole?

This emphasis on absorption capacity has been taken further through the notion of planetary boundaries introduced through the IGBP. Such planetary boundaries have to be understood within the context of Earth system science. The term "Earth systems" is developed in a major 2004 report entitled *Global Change and the Earth System: A Planet under Pressure.*[18] It is defined by Frank Oldfield and Will Steffen in this volume in the following way:

> In the context of global change, the Earth System has come to mean the suite of interacting physical, chemical, and biological global-scale cycles (often called biogeochemical cycles) and energy fluxes which provide the conditions necessary for life on the planet.[19]

On this basis, some leading scientists have identified nine planetary boundaries that define "a safe operating space for humanity based on the intrinsic biophysical processes that regulate the stability of the Earth system."[20] The assumption is that these boundaries describe a state of the Earth system that does not risk destabilizing the Holocene epoch within which human civilizations emerged.[21] They argue that humanity has faced environmental constraints throughout its history at local and regional levels but that constraints at the planetary level also have to be recognized and that the magnitude of the challenge is now vastly different.[22] As the leading author, Johan Rockström, observes, the Anthropocene raises a new question: "What are the non-negotiable planetary preconditions that humanity needs to respect in order to avoid the risk of deleterious or even catastrophic environmental change at continental to global scales?"[23]

These boundaries have been described in a 2009 report with updated data published in 2015.[24] The nine boundaries focus on climate change, biosphere integrity (rate of biodiversity loss), stratospheric ozone depletion, ocean acidification, biogeochemical flows (especially phosphorous and nitrogen), land-system change, freshwater use, atmospheric aerosol loading, and the introduction of novel entities (chemical pollution with reference to over 100,000 chemicals used industrially). All these boundaries interact with each other, but the first two are regarded as "core" since the climate system is a manifestation of the amount, distribution, and net balance of energy at Earth's surface, while the biosphere regulates material and energy flows in the Earth system and determines its resilience to abrupt and gradual change.[25] Both of these have the potential to drive the Earth system into a new state should they be transgressed substantially and persistently. The authors argue that "transgressing one or more planetary boundaries may be deleterious or even catastrophic due to the risk of crossing thresholds that will trigger non-linear, abrupt environmental change within continental- to planetary-scale systems."[26] They seek to quantify such boundaries to circumscribe what the safe operating space may be and on that basis seek to indicate a zone of uncertainty with increasing levels of risk beyond such a boundary and dangerous levels with high risks of serious impacts. The findings of the group suggest that nitrogen and phosphorous flows, together with genetic diversity loss already pose high risks, while indicators for climate change and land-system change are in a zone of uncertainty with increasing levels of risk.[27]

The authors of the *Global Change and the Earth System* recognize the policy implications of the description of such planetary boundaries. These are based on the guiding precautionary principle which is then applied in the notion of guardrails, namely, a form of management that will avoid "abrupt and irreversible changes from which human societies would find it difficult or impossible to recover."[28] This calls into question the widely used three pillars approach to sustainability that allows for trade-offs between economic, social, and environmental aspects. The report explains the complexity of this, given concerns over equity both within the current human populations and between generations.[29]

In a report entitled "Trajectories of the Earth System in the Anthropocene," Will Steffen and colleagues spell out the implications of a failure to heed such planetary boundaries:

> Incremental linear changes to the present socioeconomic system are not enough to stabilize the Earth System. Widespread, rapid, and fundamental transformations will likely be required to reduce the risk of crossing the threshold and locking in the Hothouse Earth pathway; these include changes in behavior, technology and innovation, governance, and values.[30]
>
> The Stabilized Earth trajectory requires deliberate management of humanity's relationship with the rest of the Earth System if the world is to avoid crossing a planetary threshold. We suggest that a deep transformation based on a fundamental reorientation of human values, equity, behavior, institutions, economies, and technologies is required.[31]

A final shift in the meaning of the term "sustainability" may be associated with the 15th Conference of the Parties in Copenhagen (2009) when it became abundantly clear that economic and political changes will also require social transformation, that is, changes in the hearts and minds, and especially the cultural patterns of the global consumer class. To change patterns of production require policy changes and therefore political will, which depends on voter support. This begs questions about theories and mechanisms for social change.[32] Are human societies capable of changing in order to avoid imminent environmental disaster and the subsequent collapse of such societies? How should social tipping points be understood that would elicit large-scale behavioral change? Some would argue that the climate has always been subject to change and that the human species has demonstrated a remarkable adaptability in order to survive such changes. This may be true but would underestimate the role of an unprecedented climate stability in the Holocene that has enabled agriculture and allowed civilizations to flourish. If the human species may be able to survive dramatic changes in the Earth's climate, the same may not be true for civilized societies. If mere survival is sufficient, then many of the most vulnerable people would not survive such changes.[33]

This also prompts questions around what it is that is worth sustaining, that is, what institutions and forms of culture are important enough to maintain even though they may contribute to excessive waste (e.g., a high carbon footprint). One may raise questions of this nature regarding tourism, automobiles, telecommunication systems, mega-sports events, universities, swimming pools, suburban gardens, and so on. Each of these can be maintained if less waste is produced somewhere else in the system so that the system as a whole can still absorb such waste. For that, some sacrifices have to be made somewhere. At an individual level this is easily understood, namely, to allocate more financial resources for a birthday party than for a normal family meal, as long as the total costs remain within the family's annual budget. We, rich and poor alike, readily save some funds for the special occasion. The problem here is that the costs of a large carbon footprint have to be carried unwillingly by those who have not been part of the decision-making processes, including nonhuman animals. Since the carbon footprint of some (countries) is excessive and since this pattern is being (carbon) copied for the sake of "development," there seems to be a widespread global unwillingness to stay within a carbon budget (cumulative carbon emissions) that would be sustainable, that is, one where carbon emissions can be fully absorbed through photosynthesis alone (and not through the atmosphere or the oceans). Put proverbially, the family's annual budget may show a deficit in a particular year and even over a decade if there are some reserves, but at some point trouble looms.

The distortions, inequalities, and injustices that are tolerated in socioeconomic systems can be maintained for a while, but if these are exacerbated, this will inevitably cause social instability and unrest, as is abundantly clear in contemporary South Africa.[34] In this way economic inequality, a globalized manifestation of apartheid disguising its racist roots,[35] must be regarded as a major threat to sustainability.[36] This may rightly be called the Anthropocene conundrum: how to equalize resource consumption across the world within sustainable limits, without assuming that the middle class (and the poor!) need to copy the lifestyles of the consumer class. This conundrum can only be resolved through environmental collapse or through globally coordinated efforts toward economic equality.[37] Either way, this suggests that sustainability cannot be the only relevant norm and has to be complemented by adaptability, equity, justice, and peace. Ironically, the final shift in discourse on sustainability is therefore a move away from an emphasis on sustainability (only).

This conundrum becomes paradoxical: although sustainability may be regarded in minimalist terms as a norm for prudence ("do not undermine survival"), it offers a comprehensive critique of the global economy. And, as Willis Jenkins observes, "In the debate over what societies should sustain and why, we are forced to reflect on what sustains humanity and why."[38] It is not the environment that is unsustainable; it is business as usual that cannot be sustained by the biophysical environment.[39]

To summarize: This diagnosis of the underlying problem of what is wrong with the world in the Anthropocene is that human habits, at least the lifestyles adopted by the so-called consumer class globally have become unsustainable. The Anthropocene records the collective impact of such unsustainable habits not only geologically but also on the Earth system as such. If this is an accurate diagnosis, what would be the prescribed remedy?

In isolation one may argue that there is nothing intrinsically suspect with having children, increasing one's levels of consumption (which is necessary for the hungry and destitute), and the use of sophisticated technology (which has spread rapidly throughout human history). At what point, then, do things go wrong? Is the problem merely that the impact of P × A × T is inadvertent so that it is the planetary scale of such habits that poses the danger? If so, the real dilemma is one of intentionality, namely, the inability to take collective responsibility for what is happening around us—much like a traffic jam that is not caused by any one person and cannot be resolved through individual agency.[40]

Some would suggest that the problem is indeed unprecedented precisely due to the inadvertent nature of climate change or due to intrinsic uncertainties over the long-term future. This is illustrated by the long-lasting impact of carbon emissions. This makes climate change unlike other environmental concerns that are typically marked by their immediacy.[41] It is something that is hard to notice and difficult to engage with given the long time-span involved. Naomi Klein captures the contrast between the threat of nuclear war and that posed by climate change starkly:

> The only historical precedent for a crisis of this depth and scale was the Cold War fear that we are heading toward nuclear holocaust, which would have made much of the planet uninhabitable. But that was (and remains) a threat; a slim possibility, should geopolitics spiral out of control. The vast majority of nuclear

> scientists never told us that we were almost certainly going to put our civilization in peril if we kept going about our daily lives as usual, doing exactly what we were already doing, which is what climate scientists have been telling us for years.[42]

Again, *if* the problem is indeed accurately described in terms of the diagnosis of population, consumption, and technological sophistication, where would a remedy then be found? In terms of the formula I=PAT, three remedies may obviously be prescribed: reduce the human population drastically (through draconic birth control, military extermination, induced diseases?), curb consumerist behavior (through a deliberate collapse of markets or war-time rationing?), or find new technologies that would lead to a lower carbon footprint, which is the hope of those who cannot envision ways forward in the other two areas. If the first two are politically and culturally unpalatable, then we must put all our trust in technological innovation.

At least it seems clear that what is needed is a radical transformation of the energy basis of the global economy within arguably five or six decades since 1990 when a widespread awareness of the danger posed by climate change first emerged. The first three of these decades were by far the most significant, but carbon emissions have continued to increase over this period, with no signs yet of a turning point.[43] In fact, the most recent Global Carbon Report states that "Fossil CO_2 emissions will likely be 62% higher in 2019 than the year of the 1st IPCC report in 1990."[44] Moreover, while additional sources have been added to the energy mix, no one source has been phased out. The proportional contribution of coal decreased in recent decades, but in tonnage more coal is used now than ever before.[45]

Again, how would such an economic transformation be achieved? Some Green activists suggest a drastic reduction in the use of energy within industrialized countries (instead of merely replacing the sources of energy), a spirit of communitarian austerity and sacrifice, of a return to war-time rationing, with a "carbon army" patrolling the streets. Others jeer dismissively at such a "ridiculous" option as "unworkable." Mark Lynas is at least honest: "The insistence that people must give up cars, live in colder houses, or vacation close to home is obviously a losing strategy, yet the flawed assumption that ecological limits like climate change mean that people must limit their own aspirations and lifestyles is central to mainstream Green thinking."[46] Instead, he believes, the key to finding solutions lies in innovation and aspiration. What is needed is not more moralizing but more money, to "pay the way," for example, by valuing natural capital, by putting a price on ecosystem services.[47] He therefore maintains that "Global warming is not about over-consumption, morality, ideology or capitalism." Instead, he regards it as a technical problem, "amenable to a largely technical solution, albeit one driven by politics."[48] The remedy needs to be found with technology (T) only, not with population (P) or consummation (A).

Either way, few would seriously suggest that an appropriate remedy could be found at the level of "habits of the heart," as if personal decisions to lower birth rates or reducing the household use of energy (and the embodied energy of products) can possibly address the underlying problems associated with the Anthropocene. Moral purification may serve as a way of conscientizing oneself at an individual level,[49] but unless this is widely adopted (which seems unlikely), it is not going to make the required difference. Reducing one's carbon footprint by 10 percent is as hard as making a 10 percent cut to the family budget or to staff salaries. Only a few romantics would suggest that only low levels of technology should be employed, while some optimists would hope that a miracle technological solution could make the decisive difference. How, then, is the underlying problem to be addressed? It is pertinent to recognize that there is a multilayered problem that cannot be addressed only at one level, certainly not only at the individual level. Let me illustrate this with reference to climate change discourse:

A MULTILAYERED PROBLEM

It seems clear that climate change poses scientific, technological, financial, economic, and political challenges. It makes a difference on how the problem is framed as appropriate solutions typically follow from that. As Dale Jamieson also observes,

> If the problem is fundamentally one of global governance, then new agreements and institutions are needed. If the problem is market failure, then carbon taxes or a cap and trade system is what is required. If the problem is primarily technological failure, then we need an Apollo program for clean energy or perhaps geoengineering. If climate change is just the latest way for the global rich to exploit the global poor, then the time has come for a global struggle for justice.[50]

Some climate denialists would focus on inherent *scientific* uncertainties, but there is ample scientific evidence available and sufficient scientific consensus through the work of the IPCC to suggest that this is not where the main problem lies. As Dale Jamieson observes, "Scientific ignorance abounds, but it is not the sole or even the most important reason that we have failed to act on climate change."[51]

Some suggest that the problem is *technological* or even that climate change is an engineering problem that requires engineering solutions,[52] but it seems clear that the required technologies are already available to transform the energy basis of the global economy to sustainable alternatives—to be supplemented by nuclear energy as some would insist and others would dispute. It may be true that carbon sequestration and energy storage technologies need further development, but it seems that pricing these alternatives pose a bigger problem than technical feasibility. There are some who would argue that it is not the lack of technology that is the problem but the use of existing technologies. Others would say that it is the blind faith in technology that is the underlying problem.[53] Yet it seems clear that while technology will not solve the problem, the environmental destruction caused by an industrialized economy also cannot be addressed without the use of technology. In Bruno Latour's assessment, we need to love, that is, care for the monsters (technologies) that we have created instead of blaming the monsters for causing the problem.[54]

Is the problem then *financial*? This does not seem to be the case. As the Stern report among others indicates, the costs of transforming the energy basis of the global economy are not exorbitantly high. The costs of inaction (regarded as an "externality") will eventually far outweigh the costs of funding the required transformation. Indeed, a greener economy may provide lucrative opportunities for business to some. Is the real problem then of an *economic* nature? Is the problem perhaps one of economic policies and the failure of late-modern, industrialized, neoliberal capitalism which is premised upon infinite economic growth within a finite planet? This leads to both economic inequalities and ecological destruction that have to be addressed through "development" and the allocation of sufficient finances to change the energy base of the global economy without disturbing the growth paradigm. In response, others suggest a "new economics" that would not only produce wealth (the strength of capitalism) and redistribute wealth (the strength of socialism) but will redefine how wealth is understood.

Some suggest that climate change is a *political* problem that can best be addressed through negotiations, balancing global power blocks, formulating treaties, and mustering political will. International peace is the key to addressing climate change. Others caution that justice, including climate justice, is the key to peace. What seems necessary, then, is political decisions to regulate the market in such a way that the necessary steps will become attractive to move away from fossil fuels and toward alternative sources of energy. If so, there is a need for stronger global institutions to overcome the paradox of an integrated global economy and a deeply divided global politics.[55] But then the twenty-five meetings of the Conference of the Parties thus far and the huge efforts associated with that indicate that there is indeed sufficient political consensus that this is an urgent challenge, while the Paris Agreement of 2015 at least provides some international parameters within which the matter can be taken forward. If so, why does it seem so difficult to stem the tide of carbon emissions and to move the global economy toward a more sustainable future? Is the main problem, as Jeffrey Sachs reiterates, one of implementing global cooperation and not the availability of reasonable and affordable solutions?[56]

What are the underlying reasons for political inertia? The basic problem is clear enough: The problem is deeply rooted given the energy basis of the current global economic order. There are unequal responsibilities and vulnerabilities of specific countries. Given the lag effect of carbon in the atmosphere and the delayed impact of carbon emissions, there are inherent uncertainties in understanding a changing climate, which is inherently dynamic with many feedback loops. Whereas political caution often implies a postponing of decisive moves, a wait-and-see policy may well mean waiting until it is too late in the case of climate change. Moreover, the Earth's climate system is inherently chaotic, sensitively balanced, and threshold-laden, and this allows for abrupt change with catastrophic consequences. The tendency is to treat climate change in gradualist terms where change is linear and incremental. Democratic decision-making processes are not conducive to planning over the very long term. Neither political institutions nor political theories seem to be able to cope with such a volatile problem. Any mitigation efforts would risk economic stability.[57] It is exceptionally hard to balance the economic interests of current generations against the putative well-being of future generations, given the cost-benefit analysis that is typically employed. Especially two factors lead to widely diverging results in such cost-benefit analyses, namely, the so-called discount rate to weigh future costs and benefits against present costs and benefits and the time frame (50, 500, or 5,000 years) employed. It is also questionable whether precise weights can be assigned to some benefits and costs. How does one weigh a beautiful view, friendship, or the lives of great-grandchildren? Can fine-grained calculations be made on the basis of such assigned weights?[58] Moreover, this considers the interests of other forms of life only insofar as these are relevant to human interests. The sheer complexity of the problem seems to offer a convenient excuse for further procrastination.

Many analysts would suggest that the problem will remain intransigent unless its cultural, ethical, and spiritual dimensions are recognized. The danger is that economic analysis and policy making can render ethics redundant. As the report on *Global Change and the Earth System* recognizes, advances in technology and management are essential, but on their own will not be enough. The report mentions the sheer volume of individual actions, within a still increasing human population, that drive global change and states that "alterations in societal values and individual behavior, particularly with regard to consumption, may also be required to slow or reverse current trends." Indeed, "humanity is already overshooting the capacity of Earth's biological systems to provide resources and to absorb effluents indefinitely."[59]

All humans therefore cannot adopt the material standards of living adopted by the consumer class globally.

Let me therefore comment on each of these cultural, ethical, and spiritual dimensions.

A CULTURAL PROBLEM?

It seems obvious that policy makers and politicians rely on voter support so that they would not be inclined to implement drastic changes that would alienate themselves from voters and their interests. Ecological suicide over the long term seems preferable to political suicide in the medium term when the next election looms. In democratic countries, political solutions remain dependent upon the will of the majority, albeit that the will of the majority can be bended by the corporate media. The question is what kind of world the voters want to see and whether that is sustainable. The answer may well be that "our" (American) way of life is not negotiable, as the then president Bush famously stated after the Rio Summit in 1992. One may suggest that the underlying problem around climate is less a problem of know-what or know-how than of know-why and know-wherefore.[60] The environmental crisis that we have to face is not primarily an ecological crisis but a deeper, *cultural* crisis in nature and wholly of nature, running fully against it.[61] It is a pathological sign of cultural failure and bankruptcy. It indicates that the values underlying the dominant economic and cultural practices in the world have become bankrupt. The problems may result mainly from economic production, coupled with planned obsolescence and artificially stimulated by advertising industry, but this cannot be separated from patterns of consumption. If so, many would suggest, the underlying problem is related to the consumerist culture. The critique of consumerism, including concepts such as "conspicuous consumption, the "tragedy of waste" and the "affluent society," is a vast and complex theme that cannot be reviewed here.[62] It may suffice to recall my own earlier critique of consumerism in which I identified six points of entry for a Christian critique of consumerism, that is,

- An ecological critique: The consumer society which whips up a "hurricane of desire" is not sustainable;
- An economic critique: Consumerism exacerbates economic inequalities and injustices;
- A social and psychological critique: The affluent have become the victims of their own desires; consumerism functions like an addiction,[63] a neurosis;
- A virtue ethics critique: Consumerism undermines virtue (frugality) and breeds vice (greed);[64]

- A hermeneutical critique: The consumer society encourages commercialized cultural and religious practices; and
- A theological critique: Consumerism amounts to idolatry.[65]

If the problem is indeed of a cultural nature, if technological solutions to social problems are recognized as reductionist, where can the remedy then be found? How can this incapacity to escape from the lure of consumerism be explained? This becomes a sociological or psychological question, namely, what theories, mechanisms, and strategies of change would persuade voters to demand appropriate steps from politicians? This requires further inquiry from behavioral studies, but my sense is that this will not yield promising results either. All our most trusted theories on social change seem to prove helpless to encourage the kind of cultural change that is necessary in a changing climate. Why is this the case?

- The underlying problem is clearly not just a lack of *information* or planning. Knowing what is right does not always translate into doing what is right. We do respond to available information and education but that is necessarily a slow process. Moreover, we also have to take into account that those with the highest carbon footprint are typically also among the better educated. Even those of us who are well acquainted with scenarios around climate change find it difficult to curb our carbon footprint to anything approximating sustainable levels (let us say to 2 tons of CO_{2e} per person per year). Changes are possible but only when attractive alternatives become available.
- If education fails, *marketing* provides another trusted strategy for social change. Information on a cheaper or better product or the demonstration of alternatives can prove to be very effective as long as we can recognize the positive outcomes for our lives. However, marketing works much better if the product on offer does not require from one to scale down. The problem is that marketing sends mixed messages and is better at enticing consumerism than at guiding frugal decision making.
- If marketing fails, effective *management* and innovative leadership may help to steer an organization or society through a crisis. However, the available theories of management have scarcely prepared us to address a problem that is global in scope, that will become truly ominous only in a few decades, and that will indeed require a collective, coordinated effort from at least all industrialized and industrializing countries. Indeed, climate change is the "largest collective action problem that humanity has ever faced."[66] What will be required is a form of adaptive management at an unprecedented scale.[67]

- If management fails, *punishment* offers another, no longer widely promoted option. People are typically willing to change if they recognize that not doing so will have negative implications for them. This strategy may yield some results, for example, through "the polluter pays" principle and through carbon taxes. However, there are scarcely analogies in history where the powerful throughout an empire will collectively accept punishing taxes introduced by themselves in order to effect the required change. Even then, the question is whether such taxes will yield the required 80 percent reduction in greenhouse emissions in industrialized countries. The unpalatable truth is that greenhouse emissions have been rising steadily since 1990, also in industrialized societies. The question remains whether our proposed policies are commensurate with the gravity and the global scale of the problem.
- If punishment fails, one can still call on a sense of *responsibility*. People are indeed willing to accept responsibilities on the basis of a sense of identity and their role responsibilities in a particular context. Most of us are more than willing to do our duties, to pay our TV licenses simply because it is the right thing to do, as an SABC advertisement had it. Such a sense of responsibility is fostered through moral formation which is widely recognized in virtue theory, theories of personal and social development, and theories of moral education.[68] The emphasis here is on socialization and internalization. This would prepare people to handle social change psychologically, but are hardly aimed at social transformation. What is required here, though, is (rapid) social change, not an affirmation of cultural identity. It is possible to extend such a call for responsibility, for example, toward citizenship, future generations, technology, or nature, but the broader the scope, the vaguer the specific sense of responsibility seems to become. It would be a symptom of megalomania to accept responsibility for what lies beyond one's locus of control.[69]
- If calling on a sense of responsibility fails, *prophetic warnings* and threats may be considered. This is the most trusted strategy in the media. Apocalyptic images abound in scientific reports, science fiction movies, cartoons, and heavy metal rock music alike. We human beings seem to derive pleasure from such images. We are thrilled by thrillers. One may even talk about "climate porn." However, people seldom respond to prophetic threats in an appropriate manner. Apocalyptic images prompt anxieties and fear more often paralyzes than not. It seldom promotes responsible decision making.
- A final strategy to introduce social change is to speak to peoples' desires, their *dreams*, and aspirations for the future. People are willing to endure hardships (like studying toward a degree) if the reward is attractive enough. We are quite willing to save in order to purchase something desirable. Such

dreams for the future may be based on a vision for the good society. In South Africa, many were willing to fight for justice even where this may not have been in their own immediate interests. However, there seems to be a deeper problem here, namely, that people's dreams and aspirations are being shaped more decisively by a culture of consumerism than by the belief that "a different world is possible." Indeed, the content of such dreams matters because they can become self-fulfilling prophecies.[70]

There are ecomodernists and pragmatists who believe that none of these tactics will work because the message is wrong. We need to relinquish nostalgic attachments to a nature untouched by humans. Instead of negative messaging, they argue that positive, aspirational messages are needed to design and construct a planet that will allow human communities to flourish on the basis of the progress made since the advent of modernity through industrialization, urbanization, infrastructure, and technological development *and* to create sufficient room for wildlife to flourish. Erle Ellis articulates this critique of much of contemporary environmentalism quite well when he states:

> Ironically, though, stories of environmental doom have come to dominate the discourse today. Narratives warning of a "sixth mass extinction" and many other environmental crises do have a substantial scientific basis and have helped to put serious concerns about biodiversity loss on the map. Yet such negative messaging has also been shown to be not only dismal and depressing but also disempowering, off-putting, and generally unsustainable. Most importantly, visions of future catastrophe are unlikely to motivate the vast majority of people whose lives turn on their own daily struggles, no matter how popular such narratives might be among some educated elites in developed nations. . . . There is increasing evidence to suggest that people respond more actively to positive, aspirational messages that empower them to act toward better outcomes.[71]

There is a twofold assumption in such positions, namely (a) that sustained economic growth (including an expansion in energy use) is compatible with ecological sustainability and (b) that humans have it within them, through aspirations, ingenuity, and wise management, to overcome whatever hurdles to such progress may be encountered. If these assumptions cannot be sustained, then this would amount to a shallow diagnosis of the underlying problem. I will return to debates on limits to economic growth below. For now, it may be noted that the liberal (Pelagian) optimism of such a position may be contrasted with the more starkly realistic acknowledgment of "total depravity" in the wake of Augustine and Calvin.

A PSYCHOLOGICAL PROBLEM?

Climate change may also be regarded as a psychological problem. There are by now ample studies that explore issues such as climate denialism (and its associated defense mechanisms to avoid the truth), climate anxiety, tacit assumptions that someone will come up with technological solutions, paralyzing apocalyptic fears, the crippling effect of recognizing that personal efforts will hardly make a difference, and the impact of inherent uncertainties over what the future will hold. Each of these factors tends to undermine appropriate responses at the personal level but clearly also shape decision-making processes at the corporate and political levels. Moreover, many scientists have observed that the biggest factor in uncertainties over climate change scenarios is not related to physical variables but over social and political responses that are hard to envisage.

It is important to understand how psychological factors inhibit appropriate responses. Panu Pihkala observes that apathy regarding climate change is often a manifestation of eco-anxiety.[72] In response, people have developed psychological and social defense mechanisms, including denial and "socially constructed silence."[73] Apathy in the face of climate change does not indicate, as is often assumed, that "people care *too* little or not at all, but rather that they care too much."[74] This leaves the morally sensitive ashamed of "never rebelling enough against the ubiquitous degradation of all that lives."[75] Dale Jamieson argues that evolution did not design humans to solve or even recognize the kind of problem that climate change is. We have a predisposition to respond to dramatic movements by middle-sized objects that can be visually perceived. Instead, the impact of climate change is typically invisible because it is geographically distant, statistically averaged, and temporally delayed given the lag effect of carbon in the atmosphere. Jamieson captures the problem vividly: "While we can be very responsive to individual victims, we have difficulty empathizing with statistical victims. We mobilize huge resources around highly publicized cases of little girls falling into wells while we do comparatively little to save children when they are the invisible victims of policy choices."[76] Even when we believe climate scientists telling us that the world is warming, we cannot experience global warming since we cannot be everywhere at once. We can experience weather changes but these cannot be directly correlated to climate change. We respond best to sensory impulses and not to abstract thoughts so that if we cannot sense climate change, we find it hard to act in response to that.[77]

Such psychological factors are obviously crucial but by itself it can also undermine human agency, including the agency of the very victims of climate change. The danger is that the problem becomes pathologized so that culprits are treated as victims of forces beyond their control. If the diagnosis is of a psychological nature, then some medication or psychotherapy is required without accepting personal or collective responsibility. If human evolution has ill-prepared us to deal with climate change, then it seems that only biotechnology or genetic engineering could offer a way forward. Instead, it is necessary to also confront the moral and indeed the spiritual nature of the problem.

A MORAL OR AN ETHICAL PROBLEM?

This analysis of the cultural problem of unsustainable "habits of the heart" points in the direction of the deeper *moral* and indeed spiritual nature of the problem. That there is an underlying moral problem is widely recognized given the injustices associated with the impact of climate change—those who contributed least to historical carbon emission will be affected disproportionally. Among many other sources that make the same point, let me quote from a document produced within the context of the South African Council of Churches:

> Thus far climate change has resulted mainly from the carbon footprint of the consumer class. However the attempts of others to copy this lifestyle will worsen the problem in years to come. Climate change is therefore deeply related to the skewed distribution of wealth. It is quite understandable that the poor, as well as those in the so-called middle class, would desire to share in the wealth that is so visibly portrayed in our society and through the media. This merely illustrates that the lifestyles of people in the consumer class are unsustainable since such lifestyles cannot be copied by all. Moreover, in years to come those of us with low carbon emissions will have to suffer the consequences of the economic activities of those with higher emissions.
>
> The levels of consumption enjoyed by the affluent (in South Africa) therefore raise serious questions of global justice. Such consumption levels can only be sustained at the expense of others—the poor, coming generations and other living organisms. It would simply be impossible for the planet's entire human population to replicate the lifestyle of the affluent center. The solution cannot imply a system of consumer apartheid that upholds affluent binge habits but denies the poor a decent standard of living. The affluent seemed to have wreaked environmental havoc so that they might attain a comfortable and healthy lifestyle. They are clearly not in the position to caution others not to seek a comparable standard of living, giving as reason that it would jeopardize ecological sustainability.

> It should therefore be abundantly clear that climate change raises issues of justice concerning a fair distribution of wealth, opportunities, responsibilities and also of the global costs for adaptation to the impact of climate change.[78]

The environmental justice movement that emerged all around the world, notably in India, in the United States, in the Pacific, and also in (South) Africa, has rightly demanded attention to the cases where the many environmental sins[79] of the powerful and affluent have been visited upon the poor, marginalized, minority groups and Indigenous peoples. None of these have gone away in the Anthropocene. While the impact of toxic pollution (the focus of calls to address environmental racism) has remained relatively local and traceable to polluters (who do not pay), the livelihood of numerous vulnerable communities will be affected more indirectly by a changing climate, dwindling fish stocks, and a rapid loss of biodiversity that will destabilize ecosystems. While the affluent may be able to prioritize their survival needs, this would not be possible for hundreds of millions of migrants and environmental refugees who will become displaced and will need to relocate. Rick Potts rightly comments that "Our shared evolutionary origin implies deep genetic affinity among all people," but then asks, "can this kinship and the prospect of reciprocity defeat the urges that divide us and magnify hostility?"[80]

If such a diagnosis of an underlying moral problem is accurate, may the remedy (hardly an appropriate term here) also be found in this direction? Would one speak of a moral crisis demanding moral responses? One needs to speak of an underlying moral problem carefully as this may sound like yet another call for moral rearmament or for another wave of moral regeneration.[81] If morality is reduced to introspection, examining the individual conscience, then such an emphasis may heap a sense of guilt on the sincere without making any real difference.[82] Climate change does call for moral judgment in terms of moral standards such as justice, participatory decision making (allowing a say to those most affected) and, of course, ecological sustainability. Such moral standards have already been articulated in 1975 at the Nairobi Assembly of the World Council of Churches with its notion of "towards a Just, Participatory and Sustainable Society." The Vancouver Assembly in 1983 then initiated the Conciliar process on "Justice, Peace and the Integrity of Creation," culminating in the World Convocation on JPIC in Seoul in 1990. Climate change requires attention to all three underlying moral problems, namely, economic inequalities and injustice, conflict at different levels, and ecological destruction—in relation to each other. These moral standards are also expressed in the sixteen principles of the Earth Charter and in many other documents from religious forums such as the Parliament of the World's Religions.

A reminder of such moral standards may be helpful but still does not seem to address the root of the underlying problem. Even if climate change poses the "perfect moral storm," as Stephen Gardiner suggests,[83] this does not imply that this can be resolved through ethics as theoretical, second-order reflection on the basis for moral judgments. A lack of ethical clarification does not appear to be the stumbling block even though there is an ongoing need to reflect on the relation between scientific facts and social values in climate change discourse.[84] Is our main problem that we do not recognize these standards or cannot gain clarity on conflicting moral imperatives? Is moral persuasion potent enough to induce people to make major sacrifices that go against their more immediate interests? This seems unlikely, if only because there seems to be widespread consensus on such moral norms, also in international decision-making processes, for example, with reference to the notion of a common but differentiated responsibility. Or is the problem that we do recognize but find ourselves unable to follow such standards? If so, this is a problem that has to be addressed through moral formation and not merely through more information or even ethical deliberation and decision-making processes. It is not simply a matter of agreeing with a memorandum spelling out some common values either. It would not help to blame the directors of oil companies for escalating carbon emissions if they act on behalf of their shareholders and in response to consumer demand. It is therefore appropriate that there is a widespread retrieval of virtue ethics in response to ecological concerns. The problem is not merely identifying appropriate virtues such as wisdom, justice, resilience (hope), care, and temperance.[85] This problem has to be addressed through moral formation in moral communities, which is a slow task, as any parent would realize.

The danger is that prophetic critiques of the global economic order and of climate negotiations can merely result in judgmental moral judgments. Then the prophet may merely shake her head or shrug his shoulders wondering why on Earth people are not doing what is obviously the right thing to do. Others have recognized that more is at stake, namely, the need for moral leadership, moral courage, moral vision, and moral imagination. The task of a prophet is not merely to denounce evil and injustices but also to inspire people with an attractive vision for the future, or a way out of the current dilemma, helping them to see what the next appropriate step is (although prophets are often not too bothered about implementing such steps). Martin Luther King Jr. not only had to spell out his dream for a future America but also had to suggest what ordinary citizens could do to bring that about, for example, by mustering the courage to take one's seat on a municipal bus.

To reiterate the point: the mess that we found ourselves in does raise thorny questions for ethical deliberation,[86] but it seems more a first-order problem of moral malformation than a second-order problem of ethical clarification. If so, there is not only a moral but also an underlying spiritual problem.

A SPIRITUAL PROBLEM?

It is a liberal fallacy to assume that knowledge, education, and awareness form the key to moral action.[87] It may be impossible to address environmental problems without an adequate awareness of such problems, but it is not as if such awareness was lacking until now so that scientific consensus would suffice to raise such awareness, which would then translate into technological innovation and political will. In fact, in their environmental history Christophe Bonneuil and Jean-Baptiste Fressoz conclude that "our ancestors destroyed environments in full awareness of what they were doing."[88] They argue that the problem is not the emergence of moral awareness or the lack thereof but the schizophrenic nature of modernity that regards human as products of their environments at the same time that it allows them to destroy it.[89]

If we know from science *that* we have to do something, *why* we have to that and *how* to do that (technology), but still find ourselves unable to do so, that is no longer only a moral but also a *spiritual* problem. For Christians that is a familiar problem: they know that they need to love their neighbors as themselves but often find that rather hard to do. More knowledge is not really required here. As the apostle Paul commented in Romans 7, he finds himself unable to do what he wants to do but instead does what he hates.[90] Likewise, almost all adult South Africans know what to do to avoid contracting the HI-virus, but this has not stopped the pandemic from spreading. Not only prophetic moral imperatives are needed, but also moral will and moral energy (and less patriarchal family structures!).

Where and how can such moral will be mustered? That the problem is highly complex becomes apparent with the recognition that even those who have developed an acute environmental awareness often find it difficult to translate an awareness of ecological concerns into appropriate forms of praxis. As the SACC document on climate change states,

> Those of us in the urban middle class find it difficult to adopt a lifestyle that is not harmful to the environment. We may take some modest steps to address climate change—such as reducing the use of electricity, water, transport and chemicals, while recycling and re-using resources. Such steps are highly appropriate to challenge consumerist habits and demand considerable effort and dedication. However, a guilty conscience and a 10% reduction in resource usage would not nearly be sufficient, given the scale of the problem. By contrast, those

> of us who are poor lack the resources to alter our squalid living conditions and to steer away from the (comparatively minute) environmental damage that we do cause. We naturally desire to obtain more of the wealth that we observe around us—but we can scarcely be concerned about the impact that what we desire (but do not yet have and perhaps have little hope in getting) would have.[91]

One may therefore say that collectively we human beings know that something is wrong on the basis of ample scientific evidence, recognize the scope of the problem, have the necessary technology to address the problem, can calculate that appropriate changes are still affordable and expedient, have political consensus that something has to be done but still find ourselves unable to do so. Why? Part of the problem may be self-deceit through ways of talking and thinking that prevents us from seeing the problem through an appropriate frame. The problem is not merely cognitive though, but it is also conative. Moral corruption undermines the will to do good. Corruption spreads like a virus undermining the rules by which societies abide and enfeebling the spirit to abide by such rules.[92]

This lack of moral energy is related to an inability to muster the required moral imagination, moral vision, moral courage, and moral leadership to alter the patterns of production and consumption underlying anthropogenic climate change. It is not merely an issue of psychological inertia or the freeze effect when danger looms. It is the lack of an attractive vision that leads to such inertia. In their foreword to a series of edited volumes on world religions and ecology, Mary Evelyn Tucker and John Grim comment:

> It is becoming increasingly evident that abundant scientific knowledge of the crisis is available and numerous political and economic statements have been formulated. Yet we seem to lack the political, economic, and scientific leadership to make necessary changes. Moreover, what is still lacking is the religious commitment, moral imagination, and ethical engagement to transform the environmental crisis from an issue on paper to one of effective policy, from rhetoric in print to realism in action.[93]

In the same volume, Patricia Mische adds the following:

> Science and technology alone cannot resolve ecological threats. Nor can governments or the laws they promulgate. . . . Sustaining the integrity of creation thus requires not only the external laws governments enact to deal with belligerent behavior, but also inner governance, laws internalized in our hearts and minds and the will to live by them. The need for inner governance is relevant not only to personal behavior, but also collective behavior through the economic, social and political systems we create and help maintain. Church praxis has special

> relevance for the development of inner governance and a culture of ecological responsibility. Religions carry the archetypes, symbols, meanings, values and moral codes around which people coalesce and define themselves, their sense of the sacred, and their relationships with each other and the natural world.[94]

This underlying lack of moral imagination and moral vision is also recognized from within the African context. In a *Message from African Faith Leaders to the 17th Conference of the Parties (COP17)* following a meeting at UNEP in Nairobi, June 7–8, 2011, this need for a moral vision is clearly recognized. The statement observes that "there is a profound need for a renewed moral vision for the future of humanity and indeed of all life. We debase human beings by seeing them only as economic instruments, and debase the sanctity of life by commodifying it." It then continues:

> We must realise that well-being cannot be equated with material wealth. The quality of life is not dependent on the quantity of material things or growth measured by GDP. Instead, our standard of living depends on our standard of loving and sharing. We cannot sustain a world dominated by profit-seeking, rampant consumerism and gross inequalities, and an atmosphere of competition where the powerful take advantage of the weak without caring for the well-being of every form of life. Development cannot be sustained if the affluent project themselves as examples to be copied by everyone else, and if the poor model their lifestyles on such examples.[95]

In its document on climate change the South African Council of Churches observes that "We live in a time that lacks a compelling moral vision, even though most businesses and institutions continually talk about their vision and mission."[96] The document adds that, "It is indeed a matter of moral vision. We need to envisage alternatives to the current global economic order that has caused climate change—alternatives that will be able to *generate* sufficient wealth, *distribute* such wealth more equitably and help to *redefine* our very understanding of what wealth entails. Such a vision needs to be attractive enough to motivate millions of people, to energise and mobilise action."[97]

There seems to be a lack of imagination that "a different world is possible," one that is not characterized by the now global consumerist patterns. This poses a spiritual problem that has to be addressed at its spiritual roots. The spiritual dimensions of the problem indicate the need for a form of diagnostics that is religious in depth and scope: what is *ultimately* wrong with this world? Indeed, as Pope Francis observes in his acclaimed encyclical *Laudato Si'*, "Merely technical solutions run the risk of addressing symptoms and not the more serious underlying problems."[98] And, as Ecumenical Patriarch Bartholomew I of Constantinople adds, "Climate change is much more than an issue of environmental preservation. Insofar as human-induced, it is a profoundly moral and spiritual problem. To persist in the current path of ecological destruction is not only folly. It is no less than suicidal, jeopardizing the diversity of the very earth that we inhabit, enjoy and share."[99] The spiritual nature of the underlying problem is a dominant theme in multifaith discourse on "religion and ecology" where ways in which diverse religious communities can, should, and do respond to contemporary environmental challenges are explored and assessed.[100]

It is as odd as it is remarkable that the religious depth of the problem is also recognized by secular critics. Let me mention one example, namely, the novelist Amitav Ghosh who, at the very end of his survey of literature on climate change in the humanities, regards the increasing involvement of religious groups and leaders (like Pope Francis) as "the most promising development," as a sign of hope because the formal political structures are "incapable of confronting this crisis on their own."[101] He believes that only mass movements based on religious affiliation have the ability to mobilize people in sufficient numbers. Moreover, religious traditions transcend nation-states, acknowledge intergenerational responsibilities, are able to imagine catastrophic change, are open to worldviews that do not ascribe to economistic ways of thinking and, especially are willing to accept limits and limitations because of the emphasis on the sacred and a sense of transcendence.[102]

Religious traditions cannot be understood merely as sources of moral energy though. This is not only because such traditions are often deeply ambiguous, but they are often also slow to muster such energies when called upon. Moral formation is the work of decades and centuries. It takes place in moral communities where texts, traditions, symbols, heroic figures, role models, spiritual practices, and regular exercises each play a role. A purely functionalist understanding of religion as ways of making the world a better place would underplay and indeed undermine the ultimate sources of such moral energy and therefore the virtues of resilience, resistance, and hope. Where such sources of moral energy are denied, the more realistic alternative may be to abandon all hope, to accept that catastrophic, runaway climate change is now virtually inevitable, to mourn for a lost future, to write a *Requiem for a Species*, as Australian public intellectual Clive Hamilton recommends.[103]

In any case, moral education, even if inspired by deep religious roots, is not going to make the required difference at least not on its own and not quickly enough. Why? The difficulty is that climate change cannot be easily addressed in terms of traditional theories of moral responsibility. Dale Jamieson mentions three reasons in this regard: "Apparently innocent acts can have devastating consequences, causes and harms may be diffuse, and causes and harms may be remote in space and time."[104] It is therefore not clear whether the language of responsibility (or stewardship or governance) is commensurate to the scale of the challenge of the Anthropocene. It is hard to have a sense of responsibility for the whole world or for the far-future.[105] Moreover, even those who have an acute sense of responsibility for being complicit in anthropogenic climate change (those with a large carbon footprint) do not necessarily know what options are available to them to reduce their footprint in a way commensurable to the challenge. They feel themselves trapped in societal structures determined by where they live and work (suburban homes, their use of electricity, required modes of transport, and work-related traveling) more than by individual choices. I find the portrayal of the underlying problem by Roy Scranton to the point:

> And while smart, dedicated, and thoughtful people fumble with political machinery that doesn't work, such as carbon-pricing markets, protests, and the United Nations, all of us in the Global North go about our business, driving, flying, leaving lights on, running heaters and air conditioners, eating meat, charging our devices, living unsustainable lives predicated on easy consumption. . . . The problem with the United Nations isn't that the politicians there are ignorant, hidebound, self-interested, or corrupt. The problem with our response to climate change isn't a problem with passing the right laws or finding the right price for carbon or changing people's minds or raising awareness. *Everybody already knows*. The problem is that the problem is too big. The problem is that different people want different things. The problem is that nobody has real answers. The problem is that the problem is us.[106]

To state this differently: the challenge posed to an "ethics of the Anthropocene" is that ethics would not suffice. The problem is not merely a moral one, as if we need to muster the "better angels of our nature," to encourage cooperation, sharing, and caring and to resist self-interest, apathy, and mistrust. Amitav Ghosh notes the expert opinion that an 80 or 90 percent cut in greenhouse emissions would be required in the "developed" world for the sake of fairness in a "common but differentiated responsibility." He then raises this pertinent question: "Will an abstract idea of fairness be sufficient for people to undertake cuts on this scale, especially in a world where the pursuit of self-interest is conceived of as the motor of the economy?"[107] Inversely, it is hard to imagine what ethical strategy a country like India might adopt around carbon mitigation. Should it abandon the hope of Western-style prosperity and the project of "modernization" as the reverse side of decolonization?[108] Would this not ask from the poor to make sacrifices so that the rich can continue to enjoy the fruits of their wealth? If these are not feasible strategies, then a deeper diagnosis may be required.

NOTES

1. See the widely cited paper by Paul R. Ehrlich and John P. Holdren, "Impact of Population Growth," *Science: American Association for the Advancement of Science* 171 (1971): 1212–17.

2. See Kolbert, *The Sixth Extinction*, 120.

3. See Steffen et al., *Global Change and the Earth System*, 84.

4. See Mark Swilling and Eve Annecke, *Just Transitions: Explorations of Sustainability in an Unfair World* (Cape Town: UCT Press, 2012), 96. They rightly question the implications for so-called developing countries where some economic growth is required for infrastructure (housing, transport, etc.) and poverty eradication. In such developing countries there are millions in the middle class who aspire to consume like the average Californian. They add: "Callously telling them to give up the American dream when a billion others continue to live it is how the notion of 'ecological limits' can be construed as an apology for the status quo" (98).

5. See Malm and Hornborg, "The Geology of Mankind," 65. They draw on multiple sources to substantiate the argument. The warning not to assume any direct correlation will suffice here.

6. See McNeill and Engelke, *The Great Acceleration*, 141.

7. See Ibid., 5, 209.

8. See also, from a natural science perspective, Steffen et al., *Global Change and the Earth System*, 84.

9. Steffen et al., *Global Change and the Earth System*, 134.

10. See Donella H. Meadows et al., *The Limits to Growth: A Report for the Club of Rome's Project on the Predicament of Mankind* (New York: Universe Books, 1972); also Donella H. Meadows, Dennis L. Meadows and Jorgen Randers, *Beyond the Limits: Confronting Global Collapse, Envisioning a Sustainable Future* (Post Mills: Chelsea Green, 1992).

11. I will return to the notion of limits to economic growth in the discussion below under ideological distortions in the Anthropocene.

12. The next few paragraphs borrow from my article, Ernst M. Conradie, "Is it not God's Mercy That Nourishes and Sustains Us . . . Forever? Some Theological Perspectives on Entangled Sustainabilities," *Scriptura* 116 (2017): 38–54. My analysis follows a highly perceptive early article by Larry Rasmussen in which he identifies three limits to growth, namely, economic limits (nonrenewable resources), biospheric limits (absorption capacity), and social limits (the complexity of inducing social change). One of the earliest uses of the term sustainability may be found in the 1969 International Union for Conservation of Nature document that described its goal as achieving the "highest sustainable quality of life." See Larry L. Rasmussen, "The Future Isn't What It Used to Be: 'Limits to Growth' and Christian Ethics," *Lutheran Quarterly* 27 (1975): 101–11. For a more recent overview, see Willis Jenkins, *The Future of Ethics: Sustainability, Social Justice, and Religious Creativity* (Washington: Georgetown University Press, 2013).

13. See David C. Korten, "Sustainability and the Global Economy," in *Visions of a New Earth: Religious Perspectives on Population, Consumption and Ecology*,

edited by Harold Coward, and Daniel C. Maguire (New York: State University of New York Press, 2000), 29–42 (31).

14. This widely cited definition of the World Commission on Environment and Development (the Brundtland report) reads in full: "Sustainable development is development that meets the needs of the present without compromising the ability of future generations to meet their own needs. It contains within it two key concepts: the concept of 'needs,' in particular the essential needs of the world's poor, to which overriding priority should be given; and the idea of limitations imposed by the state of technology and social organization on the environment's ability to meet present and future needs" (*Our Common Future*, Chapter 2, par. 1).

15. As Åsa Jansson observes, the notion of environmental footprint nay be useful to estimate, in a semiquantitative way, the dependence of human populations on the natural world around them. It is especially helpful to illuminate the dependence of city dwellers on ecosystems outside their immediate vicinity. However, it can provide only statistical snapshots and therefore it does not have the capability to answer questions on whether present levels of resource use are sustainable or not. See Åsa Jansson, "Ecological Footprints," in Steffen et al., *Global Change and the Earth System*, edited by Will Steffen (Berlin: Springer, 2004), 126–27.

16. For a discussion, see Ernst M. Conradie, "From Land Reform to Poo Protesting: Some Theological Reflections on the Ecological Repercussions of Economic Inequality," *Scriptura* 113 (2014): 1–16.

17. For a discussion of such concepts from within the South African context, see Steve de Gruchy, *Keeping Body and Soul Together: Reflections by Steve de Gruchy on Theology and Development* (Pietermaritzburg: Cluster Publications, 2015).

18. Steffen et al., *Global Change and the Earth System.*

19. Frank Oldfield and Will Steffen, "The Earth System," in Steffen et al., *Global Change and the Earth System* (Berlin: Springer, 2004), 7. They distinguish a number of features on this basis.

20. Will Steffen et al., "Planetary Boundaries: Guiding Human Development on a Changing Planet," *Science* 347:6223 (2015): 736–47 (737).

21. Ibid., 747.

22. Ibid., 737.

23. Johan Rockström et al., "Planetary Boundaries: Exploring the Safe Operating Space for Humanity," *Ecology and Society* 14:2 (2009): 1–32.

24. See Ibid., also Steffen et al., "Planetary Boundaries," 737.

25. Steffen et al., "Planetary Boundaries," 737.

26. Rockström, "Planetary Boundaries."

27. For details, see the 2015-report by Steffen et al., "Planetary Boundaries."

28. Steffen et al., *Global Change and the Earth System*, 294.

29. Ibid.

30. Steffen et al., "Trajectories of the Earth System," 6.

31. Ibid., 7.

32. See below, also Bob Sutcliffe, "Development after Ecology," in *The North, the South and the Environment: Ecological Constraints and the Global Economy*, edited by Vinit Bhaskar and Andrew Glyn (London: Earthscan Publications, 1995), 32–58.

33. Hamilton, *Earthmasters*, 187–90.

34. From the European context, Michel Serres puts the problem of economic inequality starkly: "Between the West and the third and fourth worlds, the divide has now reached such dimensions that, to better understand them, we need to see this divide as the depths of the abyss dug by myths: the wealthy banquet on the summit of Olympus while those condemned to death wage war hand-to-hand; obese gods communicate at leisure in every point of space without losing any time and stuff themselves with ambrosial remedies against satiety while the vast crowd of mortals loses hope of living a few hours without starvation or suffering and in peace." See Serres, *The Incandescent*, 168–69.

35. Sloterdijk, *In the World Interior of Capital*, 194.

36. See Jenkins, *The Future of Ethics*, 43–45. This is also the core argument in the South African Council of Churches' document *Climate Change – A Challenge to the Churches in South Africa* (Marshalltown: SACC, 2009).

37. See Lewis and Maslin, *The Human Planet*, 390.

38. Jenkins, *The Future of Ethics*, 169.

39. See De Gruchy, *Keeping Body and Soul Together*, 51.

40. See Grinspoon, *Earth in Human Hands*, 115. He describes four kinds of planetary changes, namely, random change, biological change, inadvertent change, and intentional change. The challenge of the Anthropocene that he welcomes is planetary changes of the fourth kind, namely, to accept the responsibility of geoengineering to overcome the impact of climate change. He argues that any global agreement to reduce carbon emissions already constitutes such intentionality so that the question is not whether the challenge is accepted, but what the content of decision making is (see 144).

41. See Hamilton, *Earthmasters*, 184.

42. Klein, *This Changes Everything*, 15.

43. See the annual report of the Global Carbon Project at http://www.globalcarbonproject.org/. Global carbon emissions have increased by an average of 1 percent per year in the 1990s and up to 3.4 percent in the 2000s except for a brief respite due to the financial crisis in 2008/2009. See also the 2019 reports by the World Meteorological Organization, at https://library.wmo.int/doc_num.php?explnum_id=9936 (accessed December 3, 2019).

44. See Slide 12, https://www.globalcarbonproject.org/carbonbudget/19/presentation.htm (accessed December 10, 2019).

45. See Bonneuil and Fressoz, *The Shock of the Anthropocene*, 101.

46. Lynas, *The God Species*, 214. See also Lynas's concluding comments on such war-time rationing and, on that basis, the possibility for carbon trading in his book *Six Degrees: Our Future on a Hotter Planet* (London et al.: Harper Perennial, 2008), 278–79.

47. Lynas, *The God Species*, 46.

48. Ibid., 66.

49. See the discussion by Kevin O'Brien in *The Violence of Climate Change: Lessons of Resistance from Nonviolent Activists* (Washington, DC: Georgetown University Press, 2017), 63–90.

50. Jamieson, *Reason in a Dark Time*, 97.

51. Ibid., 67.

52. See Hamilton, *Earthmasters*, 78.

53. See Oreskes and Conway, *Merchants of Doubt*, 261.

54. See Bruno Latour, "Love Your Monsters: Why We Must Care for Our Technologies As We Do Our Children," *Breakthrough Journal* 2 (2011). I will return to his analysis below.

55. See Jeffrey Sachs, *Common Wealth: Economics for a Crowded Planet* (New York et al.: Penguin, 2008), 5.

56. Sachs, *Common Wealth*, 11.

57. See Gardiner, *A Perfect Moral Storm*, 213–46. See also Gardiner's chapter on "Cost-benefit Paralysis," 247–98.

58. See Gardiner, *A Perfect Moral Storm*, 293.

59. Steffen et al., *Global Change and the Earth System*, 292.

60. See Larry Rasmussen, *Earth Community, Earth Ethics* (Maryknoll: Orbis, 1996), 74.

61. Rasmussen, *Earth Community, Earth Ethics*, 7.

62. For one review in the context of discourse on the Anthropocene, see Chapter 7 in Bonneuil and Fressoz, *The Shock of the Anthropocene*, 148–67.

63. One may indeed speak of carbon addictions, that is, the way in which people in industrialized societies have become reliant upon a certain way of life that require commercialized agriculture for food, urban areas free from dangerous animals (for living), electricity (for energy), and motorized transport. In the context of the Anthropocene such habits of the heart have reached a tipping point.

64. See the noteworthy study on the "greedline" by a study group of the World Council of Churches. The group investigated the possibility of defining the upper limits of consumption and wealth, for individuals and corporates, as concretely as poverty is defined. The notion of "structural greed" (in line with structural violence) suggests a rethinking of greed toward a broader definition "that encompasses human desires and structural impulses (that may overtake human behaviour) to produce and consume more than what is enough and sustainable." Such greedy structures facilitate, foster, demand, and presuppose the expression of greed among individual and groups. I will return to this aspect in the discussion of structural violence below. See Athena Peralta and Rogate Mshana (eds), *The Greed Line: Tool for a Just Economy* (Geneva: WCC, 2016), 11.

65. See Ernst M. Conradie, *Christianity and a Critique of Consumerism: A Survey of Six Points of Entry* (Wellington: Bible Media, 2009). In each case I also offered a critique of the critique in order to circumscribe the hinges upon which the debate turns.

66. Jamieson, *Reason in a Dark Time*, 61.

67. The term "adaptive management" is discussed at some length in the report by Steffen et al., *Global Change and the Earth System*, 295–97.

68. For one overview of theories of moral formation, see the study by Johannes van der Ven, *Formation of the Moral Self* (Grand Rapids: Eerdmans, 1998).

69. Sloterdijk, *In the World Interior of Capital*, 187.

70. As Erle Ellis has it, "Dreams are maps"! See Ellis, *Anthropocene*, 160.

71. Erle C. Ellis, "Nature for the People: Toward a Democratic Vision for the Biosphere," *Breakthrough Journal* 7 (2017).

72. See Panu Pihkala, "Eco-Anxiety, Tragedy, and Hope: Psychological and Spiritual Dimensions of Climate Change," *Zygon* 53:2 (2018): 545–69. My thanks to Hilda Koster for this reference.

73. Ibid., 549.

74. Ibid., 548.

75. Sloterdijk, *In the World Interior of Capital*, 147.

76. Jamieson, *Reason in a Dark Time*, 103.

77. Ibid.

78. SACC, *Climate Change – A Challenge to the Churches in South Africa*, 30.

79. Notably, the phrase "environmental sins" that are "disproportionally visited upon the poor" appears in McNeill and Engelke, *The Great Acceleration*, 193. They also refer to the socialist critique that Thomas Robert Malthus "committed the sin of failing to blame capitalist exploitation" for poverty associated with population growth (194).

80. Rick Potts, "What Will it Mean to be Human?," in *Living in the Anthropocene: Earth in the Age of Humans*, edited by W. John Kress and Jeffrey K. Stine (Washington, DC: Smithsonian Books, 2017), 27–31.

81. See Neville Richardson, "Not Another Moral Summit! Problems and Possibilities for Moral Regeneration," *Scriptura* 82 (2003): 3–14.

82. See Ghosh, *The Great Derangement*, 132–33. He fears that framing climate change as a moral issue may become a plea of last resort, when other appeals have failed to produce collective action. He adds that Gandhi's politics of moral sincerity only delayed the rush toward an all-devouring carbon-intensive economy in India.

83. Gardiner, *A Perfect Moral Storm*.

84. See the helpful discussion on facts and values in Jamieson, *Reason in a Dark Time*, 72–78. He rightly questions the "linear" model of first describing scientific facts through climate science, then determining the social impact of changes in climate and only then judging that through policy making and management. His conclusion is apt: "Whatever is true about the complex relations between facts and values, we cannot simply read what we ought to do from science, any more than we can ignore science when deciding what to do about problems such as climate change" (76).

85. This is the argument of Byron Williston in *The Anthropocene Project: Virtue in an Age of Climate Change*. In Christian ecotheology, see especially Steven Bouma-Prediger, *Earthkeeping and Character: Exploring a Christian Ecological Virtue Ethic* (Grand Rapids: Baker, 2019).

86. Such issues are ably discussed in the literature, for example, in the contributions by Gardiner, *A Perfect Moral Storm*, and Jamieson, *Reason in a Dark Time*.

87. Steven Pinker provides statistical evidence to the contrary, indicating a correlation between education and moral progress. See his *Enlightenment Now: The Care for Reason, Science, Humanism and Progress* (London: Penguin, 2018), 214–32, especially 228: "Knowledge and sound institutions lead to moral progress." By contrast, I argue that knowledge of reality may be a requirement for the realization of the good, but it is not sufficient.

88. See Bonneuil and Fressoz, *The Shock of the Anthropocene*, 196. They demonstrate this conclusion with a survey of environmental reflexivity throughout the rise of modernity.

89. See Ibid., 197.

90. See Romans 7:14-24 "I do not understand what I do. For what I want to do I do not do, but what I hate I do. . . . For I have the desire to do what is good, but I cannot carry it out. For I do not do the good I want to do, but the evil I do not want to do—this I keep on doing . . . So I find this law at work: Although I want to do good, evil is right there with me. For in my inner being I delight in God's law; but I see another law at work in me, waging war against the law of my mind and making me a prisoner of the law of sin at work within me. What a wretched man I am!" (NRSV).

91. SACC, *Climate Change – A Challenge to the Churches in South Africa*, 19.

92. See Gardiner, *A Perfect Moral Storm*, 301.

93. Mary Evelyn Tucker and John Grim, "Series Foreword," in *Christianity and Ecology: Seeking the Well-being* of *Earth and Humans*, edited by Dieter T. Hessel and Rosemary Radford Ruether (Cambridge: Harvard University Press, 2000), xv–xxxii (xix).

94. Patricia Mische, "The Integrity of Creation: Challenges and Opportunities for Praxis," in *Christianity and Ecology: Seeking the Well-being* of *Earth and Humans*, edited by Dieter T. Hessel and Rosemary Radford Ruether (Cambridge: Harvard University Press, 2000), 591–602 (592–93).

95. This communiqué was compiled jointly by 130 faith leaders representing Muslim, Christian, Hindu, African traditional, Bahá'í, and Buddhist communities from thirty countries across Africa. It was circulated electronically via email to participants and is not readily available from the internet.

96. SACC, *Climate Change – A Challenge to the Churches in South Africa*, 40.

97. Ibid., 39.

98. Pope Francis, *Laudato Si': On Care for Our Common Home* (Vatican City: Encyclical Letter, 2015), par. 144.

99. See Ecumenical Patriarch Bartholomew, Statement for the WCC Working Group on Climate Change, in *Climate Change* (Geneva: WCC, 2005), 67. A "Declaration on the Environment" signed by Patriarch Bartholomew and Pope John Paul II on June 10, 2002, expresses a similar assessment: "What is required is an act of repentance on our part and a renewed attempt to view ourselves, one another, and the world around us within the perspective of the divine design for creation. The problem is not simply economic and technological; it is moral and spiritual. A solution at the economic and technological level can be found only if we undergo, in the most radical way, an inner change of heart, which can lead to a change in lifestyle and of unsustainable patterns of consumption and production. A genuine conversion in Christ will enable us to change the way we think and act." See http://w2.vatican.va/content/john-paul-ii/en/speeches/2002/june/documents/hf_jp-ii_spe_20020610_venice-declaration.html (accessed August 17, 2016).

100. This sentiment is widely expressed in the context of the Forum on Religion and Ecology and through the work of Mary Evelyn Tucker, John Grim, and many others, typically drawing inspiration from the wisdom of Thomas Berry.

101. Ghosh, *The Great Derangement*, 159.

102. Ibid., 160–61.

103. See Hamilton, *Requiem for a Species*. Hamilton recommends despair, acceptance, and ethical action that may include civil disobedience (226). He takes advice from Pablo Casals: "The situation is hopeless; we must now take the next step," and adds that, "Finding meaning in adverse circumstances is one of the most remarkable human qualities" (222).

104. Dale Jamieson, "Ethics, Public Policy, and Global Warming," *Science, Technology, and Human Values* 17:2 (1992): 139–53. Also quoted in Antonaccio, "De-moralizing and Re-moralizing the Anthropocene," 129. In *Reason in Dark Times* Jamieson mentions the magnifying impact of technology, the spatial reach of climate change, and the complexity of carbon cycles (161–62). Together these constitute the most complex collective action problem in human history.

105. See Ten Bos, *Dwalen in het Antropoceen*, 72.

106. Scranton, *Learning to Die in the Anthropocene*, 67, 68.

107. Ghosh, *The Great Derangement*, 146.

108. Ibid., 148–49.

Chapter 4

Structural Violence in the Anthropocene?

STRUCTURAL VIOLENCE

Apartheid clearly entailed cultural "habits of the heart" that captured the ways of life of ordinary South African citizens. "Capture" is an appropriate term here given the role of colonial conquest, land grabs, slavery, subjugation, and exploitation that played itself out in what became the Union of South Africa in 1910. Culturally, this led to the habitual separation of people on the basis of race and ethnicity in order to maintain cultural identity. Such a culture of "staying apart" could be found among those who assumed their cultural superiority (white supremacy) given the symbols of ships, guns, and books, but also among those who internalized the implied inferiority psychologically and those who opted for separation as a strategy for resistance against exploitation. Such cultural separation was of course never really absolute given the ongoing economic interaction between separated groups (slavery), the psychological impact of such separation (mutual fear, suspicion, and hatred), and the sexual intercourse that took place despite whatever legislation or social taboos may have been in place, whether through rape, prostitution, or love affairs.[1]

Yet, it should be obvious that apartheid entailed far more than such cultural habits of the heart, even if that describes the daily interaction between groups separated on the basis of culture, language, race, and ethnicity. Apartheid had military, economic, and political dimensions that came together through the policies implemented by the National Party government after 1948 that may be understood as a crude radicalization of the policy of segregation adopted throughout the British Empire, certainly also in the colonies that were integrated into the Union of South Africa. Put differently, the cultural habits of the heart shaped the hearts and minds of voters who supported the

National Party government, but the implementation of apartheid policies also reinforced such a culture through the institutional structures put in place in every sphere of society. The domination in the name of differences of race and ethnicity embedded in such structures is aptly described as one form of "structural violence."

Violence (and the injustices associated with that) is not only perpetrated by individuals, groups, or larger collectives (such as armies). Individual acts of violence (and injustice) are reinforced by cultural habits. This can become embedded in social structures, in institutions and their policies, in systems of inclusion and exclusion, in societies and their implicit or explicit laws. Such violence can be sustained over longer periods of time when it becomes internalized and accepted as "normal," taken for granted as the way things are and might as well be. There are many obvious examples here: apartheid, the caste system, colonization, commercialized animal farming, heteronormativity, patriarchy, slavery.

The term structural violence is derived from the work of Johan Galtung who distinguishes between direct violence, structural violence, and cultural violence. Galtung acknowledges verbal and physical forms of direct violence that can harm body, mind, and spirit. He states, "Structural violence divides into political, repressive and economic, exploitative; supported by structural penetration, segmentation, fragmentation and marginalization. In addition, there is also the horizontal structural violence of being too tightly related, and of being too loosely or evenly unrelated. Structures can be too dominant (vertical), too tight (horizontal) too much (both/and), and too little (neither one, nor the other)."[2] Structural violence is characterized by vertical relations leading to political repression (of freedom) and economic exploitation (of well-being). Cultural violence operates through language, art, philosophy, education, and the media by preventing conscientization through penetration (conditioning the minds of others from above) and segmentation (limiting the vision of reality of those below) and by preventing mobilization through fragmentation (splitting those below away from each other) and marginalization (setting those below apart from the rest).[3] Ideology (see the next chapter) therefore plays a crucial role in cultural violence.

In the famous analysis of Dom Helder Camara, taking the debate further, such structural violence typically leads to a further spiral of oppressive, revolutionary, and repressive violence.[4] Helder Camara named these poverty, revolt, and repression and insisted that the only way to stop the spiral of violence is to have the courage to face the injustices which constitute oppressive (structural) violence.[5]

In the Anthropocene this analysis of direct and more indirect modes of domination and exploitation may be taken a step further by observing that the impact of such violence is not only on social systems but also on Earth systems. As Paul Crutzen and Will Steffen observe, the "period of the Anthropocene since 1950 stands out as the one in which human activities rapidly changes from merely influencing the global environment in some ways to dominating it in many ways."[6] There is a rupture in the interplay between the geological forces of nature, that is, between the biosphere, the atmosphere, the hydrosphere, and the geosphere, through the noosphere. It may not be appropriate to refer to earthquakes and volcanic eruptions as "violent," except anthropomorphically. However, the point of recognizing the Anthropocene is precisely that the destructive forces of nature are now influenced by the cumulative human impact. The novel element in the Anthropocene is that a nonphysical force (human decision making) is exerting destructive geological power along with other such physical forces.[7]

This takes discourse on structural violence to another level, namely, that of Earth system science so that there is indeed a new element here. Put differently, there is now a dialectic between structural violence (through economic exploitation of natural resources), the volatility of the Earth system (e.g., climatic changes), and the impact of such natural forces on those who have already been marginalized by such economic exploitation. Such marginalized groups include people in impoverished countries whose carbon emissions are well below sustainable levels but who are suffering disproportionally from a changing climate, deforestation, and overfishing, who will become increasingly vulnerable and who are likely to be or become environmental refugees. This category also includes minority groups (Indigenous peoples, descendants of slaves, migrant workers) in countries with historically higher levels of carbon emissions who are the victims of climate injustices and toxic pollution. It certainly also includes countless other forms of life whose habitat has been destroyed or that cannot adapt fast enough to a changing climate and species that will become extinct due to anthropogenic impact.

There is no need here to review the structural violence embedded in the current global economic order and its impact on poverty, increasing economic inequality, violent conflict over scarce resources (water and oil), social and psychological misery, and the destruction of ecosystems. This can fill encyclopedias and will require expertise that I cannot offer. I can merely note the critique, in widely available literature, of neoliberal capitalism, the role of international corporations, especially fossil fuel companies, trade agreements, the World Bank and the International Monetary Fund, the concentration of wealth among the super-rich, the role of banks and stock exchanges, and so on and so forth. Instead of a review of such mechanisms through which structural violence is perpetrated and maintained, I may offer three loosely related comments on the interpretation of such structural violence as a diagnosis of what is wrong in the world. These focus on (1) the indirect if not unintentional ways in which structural violence operates; (2) the appropriateness of the term "wicked problem" to capture such a diagnosis; and (3) the interplay between the Anthropocene and other social divides associated with race, class, gender, caste, culture, language, and religion.

INDIRECT AND UNINTENTIONAL?

First, discourse on structural violence raises questions of agency and hence responsibility. Can those who benefit from structural violence still be held responsible or are they in a way caught up in a system (a thread-mill) that is beyond their locus of control? In the context of debates on the Truth and Reconciliation Commission in South Africa, this has become a bone of contention. The commission focused on gross violations of human rights and considered compensation for the victims or their relatives and amnesty for perpetrators, as long as they told the truth and could explain why their actions were politically motivated. Many commentators noted that this excluded the so-called beneficiaries of apartheid, that is, those who were systematically advantaged over decades if not centuries by military conquest, political systems, economic policies (taxes and benefits), and social mechanisms (housing, schooling, health services, affirmative action).[8] Likewise, those who were disadvantaged but who were not necessarily the victims of gross human rights violations were reduced to spectators. As I argued elsewhere, this requires a far more differentiated analysis where distinctions between architects, implementers, enforcers, financers, supporters, beneficiaries, bystanders, spectators, correspondents, and so forth are recognized.[9] This requires further reflection on the interplay between individual, collective, and institutional agency and the social structures within which people, communities, and organizations are situated. Can this help to understand structural violence in the Anthropocene?

Much further reflection is required, but one contribution coming from the World Council of Churches may be helpful, namely, the report from the Greed Line Study Group entitled *The Greed Line: Tool for a Just Economy?*[10] The report addresses economic inequalities in the wake of the financial crisis of 2008. It notes that a poverty line is often employed in economic literature while excessive wealth is often discussed and measured but hardly scrutinized. The report uses the term "greed line" to explore the question at what point wealth becomes excessive and needs to be redistributed for the sake of justice, sustainability, and social stability. It counters the assumption that the production of wealth does not need to harm the poor since a rising tide can lift all boats, given the planetary limits to such growth. (There is an ebb for every flow!) The considerable details on whether such a greed line can be established need not concern us here. What is interesting is the way in which the notion of greed is extended beyond individual or institutional agency (corporate greed) to the kind of structural violence that is at stake here.

In his essay Konrad Raiser, the former WCC general secretary, refers to Galtung's notion of structural violence and defines structural greed accordingly as manifestations of greedy accumulation through structures and structural arrangements. He acknowledges that such structures are shaped by actions and ambitions of persons but insists that structural greed cannot simply be reduced to greedy actions or to the desires of individuals in positions of authority in institutions.[11] He then identifies the following characteristics of structural greed:

- It operates independently of individual greedy intentions and desires;
- It does not necessarily have the obvious characteristics of greedy actions, but is the result on institutional arrangements, processes, and actions that stand in a complex causal relationship;
- It cannot simply be attributed to the responsibility of a single person or institution;
- It brings out into the open nonsustainable social, economic, political, and ecological inequalities and imbalances that are potential sources of conflict; and
- It manifests itself in the propagation of values that serve to legitimate a culture of (individual) greed as manifested in the collective pattern of consumerism.[12]

Structural greed therefore cannot be based only on an analysis and moral critique of the motivations and intentions that shape individuals. It operates through the privatization, commodification, and accumulation of common or public goods. It has structural consequences in terms of instigating or at least tolerating poverty, unemployment, and inequality and violating ecological limits. It sets patterns that others are more or less forced to adopt and follow. It reinforces individual choices through the "self-legitimizing framework of values that sustain and promote a culture of consumerism."[13]

In a contribution to the same volume, Jung Mo Sung takes this analysis a step further by suggesting a distinction between the individual, the institutional, and the structural. He observes that a diagnosis of personal and institutional greed can be remedied by promoting generosity and cultivating compassion for others. Changing the desires of a social group or institution may be more difficult than for an individual or family, but it can still be done through leadership and (democratic) decision-making processes in which values and virtues play a role. Structural greed has a different dynamic though. Structures operate through self-organization that does not depend on the intentions of individuals or institutions. What matters is the efficiency or inefficiency of such agents within the system. He explains:

> In an economic system driven by structural greed, a person who is not greedy but is efficient in his or her work can contribute more to structural greed than a greedy person who is ineffective in his or her assigned role within the system. When we speak of structural greed, we must abandon the moral of subjective intentionality. People's intentions count for very little. If an economic system's principles are maintained, changing the intentions of people does not change the logic of structural greed.[14]

Sung acknowledges the awkwardness of the term "structural greed" since structures cannot desire as people do. This requires a modified notion of greed. Following the notion of structural sin in liberation theology, Sung argues that structural greed is a form of structural sin. He explains: "Under 'structural sin' or 'structural greed,' sin is committed regardless of the intent of the person. Sin is not committed through intentional violation of moral, religious, economic or civil laws. On the contrary, sin is committed in law enforcement, in acting according to laws."[15] The conceptual clumsiness of the term aside, Sung recognizes that this requires a critique of the dominant global economic order of capitalism (based on economic scarcity, the need for sustained economic growth, and the limitless accumulation of capital), its tacit anthropology (human self-interest forms the paradoxical key to the common good), and its underlying worldview (premised on modernity). His verdict is that the structural greed of capitalism is based on its self-organizing mechanism (the dynamics of the "free" market, freed, that is, from social goals) that devours the sources of wealth and life. Those who participate efficiently in the logic, he observes, "breed and feed, either consciously or unconsciously, willingly or unwillingly, structural greed."[16]

Can this critique of capitalism be extended to a diagnosis of what is wrong in the Anthropocene? One point seems pertinent. It may be relevant to instruct individuals on their carbon footprint and to challenge the corporate greed of the fossil fuel industry and its shareholders, but this does not yet address the structural problem of the reliance of the global economy on fossil fuels as the major source of energy—black oil and coal that functionally replaced the reliance on black slave labor.[17] Naomi Klein's conclusion in *This Changes Everything!* is to the point here. Klein argues that the global inability to curb carbon emissions is because the responses required directly challenges the reigning economic paradigm, namely, a deregulated form of capitalism. The logic that encourages the liberalization of "free" trade, the extraction of natural resources at the lowest cost, and the exploitation of cheap labor is the same as the one that exploits fossil fuels for the sake of cheap energy and factors in pollution and carbon emissions as externalities.[18] In fact, it seems easier to accept a chaotic and disastrous changing climate than to accept the prospects of altering the growth-based, profit-seeking logic of capitalism.[19] Or, in the famous quip by Fredric Jameson, it nowadays seems easier to imagine the end of the world than to imagine the end of capitalism.[20] This comes despite the conclusion of the Stern report that climate change is the symptom of the most wide-ranging market failure ever seen.[21]

Klein illustrates this with the difficulty of regulating the profitable fossil fuel industry even where alternative sources of energy become available. She explains the need for oil companies to maintain carbon reserves that can be extracted after those currently in production are depleted. The ratio should be 1:1 or else the viability of a company will become suspect among its shareholders. However, if all the available fossil fuels are burnt by the end of this century, this will lead to runaway climate change. This has prompted activist slogans such as "Keep the coal in the hole and the oil in the soil!" Klein concludes that what is necessary to address climate change is the very thing that these oil companies cannot contemplate doing without initiating their own demise. Cutting emissions radically is therefore "simply not compatible with the continued existence of one of the most profitable industries in the world."[22]

If so, this raises the question as to whether and how structures can be changed. Structures do evolve over time, but they can also be changed through deliberate decision-making processes. More often structural changes follow upon the collapse of a structure due to internal contradictions or outside pressures. One is reminded here of Jared Diamond's popular book *Collapse*, in which he describes several cases in which a local culture collapsed that could have been avoided if that culture understood their ecological situatedness better.[23] The failure to respond may be a function of a failure to anticipate the problem, to perceive the problem once it has arisen, or to succeed in attempts to address the problem. In the context of climate change discourse, the time delay between foreseen consequences and prudent decision-making processes would make such a collapse disastrous. This prompts two contrasting human responses. On the one hand, one knows that it would be irrational to wait for a tumor to emerge before stopping smoking. It is necessary to change the direction of an oil tanker timeously to avoid an iceberg. On the other hand, a renunciation of pleasure today in the face of a likely catastrophe fifty years hence is a strategy that seems doomed for failure.[24]

A WICKED PROBLEM?

Second, climate change has often been described as a "wicked problem," even a "super-wicked problem."[25] If so, what exactly is "wicked" about a wicked problem, does that apply to the Anthropocene in general (and not only to climate change) and does this concept help to deepen the diagnosis of what has gone wrong in the world as a result of structural violence?

In what ways is a "wicked problem" any different from a serious, difficult, complex, tricky, or multidimensional problem? The term wicked problem was used in a guest editorial by the influential systems theorist West Churchman in the journal *Management Science* in 1967 in response to an earlier paper by Horst Rittel in which the term was introduced.[26] Churchman emphasized the responsibility to inform managers in what ways solutions to management problems have failed to "tame" such "wicked" problems. The need to "tame" a problem seems to suggest that it would no longer show its teeth before it bites! In this context, a wicked problem is a one that is "difficult or impossible to solve because of incomplete, contradictory, and changing requirements that are often difficult to recognize."[27] There is no moral connotation here as if "wicked" means "evil"; it implies difficulty or resistance to resolution. This is the gist of Churchman's editorial, namely, to question the connotation of deception to make people think something is so when it is not. Bringing in morality, he says, is "ridiculous" for "many a management scientist" who tackles only problems that are "feasible." Where a problem proves to be too complex, Churchman advises honesty and responsibility, namely, to (reluctantly) inform managers in what respect proposed "solutions" have failed to tame a wicked problem (despite considerable costs to the company in terms of data collection and analysis). Yet he admits that irresolvable problems pose problems not only to the scientist but also to the professions of operations research and management science and recommends that these professions take themselves seriously in order to become more mature as a science.[28] The scientific pretense and the managerial ethos of his position are evident: problems are there to be solved through adequate management.

The term was further developed in a widely cited article by Horst Rittel and Melvin Webber, both urban planners at the University of California in Berkeley.[29] They also contrast such wicked problems with "tame" problems investigated by "normal science" that are more readily resolved, for example, problems in mathematics, chess, or puzzle solving. In addition, one may distinguish between convergent and divergent problems.[30] Wicked problems, however, are never solved; they are only resolved—over and over again.[31] They did not use the word "wicked" in the sense of ethically deplorable, implying malicious intent, but in the sense of being "malignant" (instead of benign), or "vicious" (as in a vicious circle), or "tricky" (like leprechauns), or "aggressive" (like a lion in contrast with the docility of a lamb).[32] They identify the following ten characteristics of a wicked problem, devoting a section to each of these:

- There is no definitive formulation of a wicked problem.
- Wicked problems have no stopping rule (to know when a problem has been resolved).
- Solutions to wicked problems are not true-or-false, but good-or-bad.
- There is no immediate and no ultimate test of a solution to a wicked problem.
- Every solution to a wicked problem is a "one-shot operation"; because there is no opportunity to learn by trial and error, every attempt counts significantly (previous attempts to address wicked problems leave traces that cannot be undone).
- Wicked problems do not have an enumerable (or an exhaustively describable) set of potential solutions, nor is there a well-described set of permissible operations that may be incorporated into the plan.
- Every wicked problem is essentially unique.
- Every wicked problem can be considered to be a symptom of another problem.
- The existence of a discrepancy representing a wicked problem can be explained in numerous ways. The choice of explanation determines the nature of the problem's resolution.
- The social planner has no right to be wrong (planners are liable for the consequences of the actions they generate).[33]

The circular nature of a wicked problem is important. The problem cannot be defined accurately because the solution depends on how the problem is framed and vice versa: the definition of the problem depends on the proposed solution. Wicked problems have uncertain boundaries because they are symptoms of other problems and vice versa in a circle of mutual implication. Moreover, the stakeholders involved adopt diverging worldviews and therefore frame the problem in different ways. For Rittel and Webber, the very attempt to find a scientific basis for social policy is bound to fail because of the nature of wicked problems. They argue that the societal problems that urban planners deal with are different from the problems encountered by engineers because planning problems are inherently "wicked."[34] In particular, they find wanting the classical systems approach developed in the military and space programs where planning can be organized in distinct phases to gather, analyze, and synthesize information in order to then work out a solution. This does not work because understanding the complexity of the problem only emerges gradually. Indeed, urban planning differs in complexity from military planning!

The system theorist Russell Ackoff, Churchman's first doctoral student and later coauthor, employs the related term "mess" to describe a set of interrelated problems where every problem interacts with other problems, while solutions to any one problem produce other, often unforeseen problems. The solution to a mess cannot be to solve each of the problems of which it is composed independently. As examples, he mentions that efforts to deal separately with specific aspects of problems such as transportation, health, crime, and education only seem to aggravate them. From his perspective, namely, with an understanding of systems as open and dynamic, what is required is planning instead of problem solving.[35] He describes such planning as participative, coordinated, integrated, continuous, and prospective. The crucial step in such planning is to "design" an idealized future for the system, beginning "from scratch." He describes this step as "fun" because no special skills are involved and tellingly adds that "Playing God always is [fun]."[36] Indeed, this seems like a managerial version of "intelligent design"! Once an idealized design has been prepared with sufficient consensus, constraints have been considered, and the design is being implemented, there is a need for continuous revision in the light of experience and feedback loops. He refers to the system for making such revisions as the "planning control system" and understands "control" as a continuous process of responding and adapting frequently and rapidly to changes. Ackoff states that the controlling (management) subsystem needs to perform four functions quickly and efficiently: (1) identify problems, (2) make decisions and plan, (3) implement and control the decisions, and (4) provide the information required for each of the first three functions.[37] He describes the "state of the art science of management" in an appendix and expresses the hope (in 1974!) that many of these functions can now be automated so that humans can be relieved of routine "drone-like" tasks and can focus on more complex and challenging tasks.[38]

It is important to understand the contexts from within which the term "wicked problems" emerged, namely, design theory and management science. The quite extensive literature on wicked problems in the economic and management sciences seeks to resolve such problems procedurally and administratively through computer-based modeling, software development, strategic "interventions" or quasi-medical (or even military) "operations" in one way or another. The assumption is that a problem should be resolved, could in principle be resolved, but that this proves to be more difficult than envisaged. In the case of technological design flaws (e.g., in an aircraft), it is clearly desirable to eliminate such wicked problems or else it may have fatal consequences. A wicked design problem cannot be solved by the application of currently available methods and therefore demand creative solutions. In business strategies, a proposal may prove to be counterproductive but cannot be first tested before it is implemented. In managing human resources, it is less easy but still desirable to avoid unintended consequences of proposed new procedures. In the case of urban planning and policy making, it is even harder to envisage the effect of proposed legislation. It is not possible to do so by trial and error since the implementation of urban planning always has long-term consequences. An engineering approach (in the form of social engineering) may actually aggravate the problem being addressed. While efficiency is an engineering requirement, it may be part of the problem in social engineering. This is graphically demonstrated by the enforcement of Group Areas in South Africa during the apartheid period.

This indicates a modernist mind-set where problems are there to be solved through rational argumentation, calculation,[39] management, and policy making. Complex social problems are analyzed in terms of their constituent parts that are then addressed in their particularity—only to find that these constituent problems are interrelated and therefore resists ready-made solutions. With such a reductionist, perhaps even technocratic logic it is no wonder, despite the use of the term "wicked," that such wicked problems are difficult to resolve since moral discernment is not even considered.

While the focus of such literature on wicked problems was on systems design at a micro- or intermediate level, the concept was soon applied to broader social and economic policy problems. It is therefore not surprising that unemployment, crime prevention, obesity, the HIV/AIDS pandemic, drug trafficking, and climate change are considered to be "wicked problems" as they do not confirm to expectations in design theory to find technical solutions to design flaws. Especially climate change is widely described as a wicked problem, typically with reference to Rittel and Webber and without further elaboration or critique.[40] Perhaps it is the circular nature of the interplay between modernity, military conquest, slavery, capitalism, science, technology, industrialization, carbon emissions, and global warming that best illustrates why this is regarded as a wicked problem.

Three further contributions to the available literature may be mentioned:

First, in a document produced by the Australian Public Service Commission climate change is specifically described as a wicked problem. Building on the work of Rittel and Webber, it lists the following characteristics of a wicked problem: they are difficult to clearly define, full of interdependencies and are multi-causal, attempts to address wicked problems often lead to unforeseen consequences, are typically not stable and constantly evolving, have no clear solutions, are socially complex, hardly ever sit conveniently within the responsibility of any one organization, involve the need to change behavior (e.g. through legislation, subsidies, fines, taxes), and seem intractable and are characterized by chronic policy failure.[41]

Second, Richard Lazarus, an environmental law scholar, relates such wicked problems to climate change by identifying three additional features.

One is that time is not costless: the longer it takes to address the problem, the harder it will be to do so. Every year that passes without a drastic reduction in global carbon emissions makes catastrophe more likely. Lazarus explains:

> As greenhouse gas emissions continue to increase, exponentially larger, and potentially more economically disruptive, emissions reductions will be necessary in the future to bring atmospheric concentrations down to desired levels. Future technological advances, therefore, would likewise have to be able to achieve those exponentially greater reductions to make up for lost time. The climate change that happens in the interim may itself cause sufficient economic disruption, for instance, by slowing growth rates, so as to make it much harder to accomplish the necessary technological innovation.[42]

Another feature of climate change is that those who are in the best position to address the problem are not only those who caused it but also those with the least immediate incentive to act within that necessary shorter time frame. The most powerful nations, especially the United States, are unlikely to embrace restrictions on their own economies and are least susceptible to demands by other nations that they do so. Lazarus regards it as a "perverse irony" that such powerful nations are also least likely to suffer from changing climates.[43]

A third feature is the absence of an institutional framework to develop, implement, and maintain the laws necessary to address climate change. Lazarus concludes that despite the UN's Framework Convention on Climate Change there is "no international law making and law enforcing institution with a jurisdictional reach and legal authority that match the scope of the problem."[44]

This analysis of wicked problems has been taken further in a third important contribution by Kelly Levin and colleagues who describe what they term "super wicked" problems, with specific reference to climate change. Four additional key features are identified: time is running out; those who cause the problem also seek to provide a solution; the central authority needed to address it is weak or nonexistent; and, partly as a result, policy responses discount the future irrationally (more or less in the same way that smokers postpone thinking about the health consequences of their habit). While the items that define a wicked problem relate to the problem itself, the aspects that define a super-wicked problem relate to the agent trying to solve it. The agent is caught up "within" the problem.[45] They observe that these four features "combine to create a policy-making 'tragedy' where traditional analytical techniques are ill equipped to identify solutions, even when it is well recognized that actions must take place soon to avoid catastrophic future impacts."[46] The gist of their contribution is to guide policy making when confronted with such wicked problems.

It is evident from this brief overview of the literature on wicked and super-wicked problems that the complexity of the underlying problem is indeed acknowledged. There seems to be some recognition, grudgingly at times, that complex social problems cannot be resolved through engineering, management, or policy making. However, there is also a refusal to even consider a moral dimension to the problem, despite the use of the term "wicked." Wicked does not even mean playfully mischievous (as in the "wicked witch of the west"), but points to frustrated attempts to perfect a design. There is no recognition that such a managerial ethos could itself become an expression of violence. It is therefore hard to see how this term can help to better understand the impact of structural violence in the Anthropocene. Ironically, this conclusion reinforces the recognition by Rittel and others that the description of a problem may well aggravate the problem. To label a problem as "wicked" does not help to resolve it, especially if this serves as a source of inspiration for more aggressive, ever-more frantic attempts to find solutions for it. This will become pertinent in the discussion below of geoengineering as a quasi-soteriological solution to the "wicked" problem of climate change. Moreover, to label a problem as "wicked" without reference to the moral dimension of the problem indicates an ideological blindfold (see the next section below). As climate scientist Mike Hulme observes, "Science is never enough to resolve problems that are cultural in origin."[47] This would not yield an adequate diagnosis of what is wrong in the Anthropocene.

Let me mimic a formulation by Slavoj Žižek. He asks: "Why are so many problems today perceived as problems of intolerance, rather than as problems of inequality, exploitation or injustice? Why is the proposed remedy tolerance, rather than emancipation, political struggle, or even armed struggle?"[48] Why, then, is a problem called "wicked" if this serves as euphemism for violence? Does wickedness not refer to something evil, something even worse than verbal, physical, structural, and cultural violence?

Yet, there is also an opposite need to see the danger of moralizing discourse on wicked problems. Consider the examples of economic inequality, violent conflict, and climate change. None of these "wicked" problems could be regarded as amoral, but they are also not obviously manifestations of wickedness in the sense of "intentionally doing acts that are morally wrong."[49] Moreover, as Mary Midgley observes in her book on *Wickedness*, it is not so easy to determine evil intentions given the problem of natural evil and its interplay with social evil. Inclinations toward assertion, aggression, anger, resentment, and fear may well be appropriate in certain circumstances if they enhance well-being and are not used toward self-destructive ends.[50] To be sure, there are many evil consequences associated with such inclinations, but by themselves emitting carbon dioxide, maintaining economic interests, engaging in conflict, and self-protection cannot be equated with doing evil.

Those who contribute disproportionally to climate change are not necessarily doing so with evil (wicked) intentions or as a result of gross negligence. In fact, they are often trying to do good: purchasing food for the family, fetching the children from school, flying for work (attending another global conference on climate change), running a hot bath for personal and family hygiene, and so forth. This applies at the individual level but also at the national level where priorities have to be juggled in order to address concerns over education, health, poverty, and the climate.

Perhaps it is best to regard evil as the consistent pursuit of something that is good in its own right (e.g., motives to secure food, self-protection, or sex), but in isolation from other goods, thus giving rise to pernicious evils. According to the tradition of sin as privation of the good (*privatio boni*), it is the perversion, recombination, and narrowing of desires that are otherwise natural.[51] Climate change is the cumulative, mostly unintended outcome of daily activities, stretched over several generations and exacerbated by corporate and political power. To reduce structural problems to moral wickedness would therefore evade the underlying difficulties.[52] Nevertheless, intentionality is not the only prerequisite to judge something to be horribly wrong, as Adolf Eichmann's famous trial suggests. In fact, evil becomes manifest not so much in the clarity of unambiguously evil intentions, but in confusing moral distinctions between good and evil.[53]

Inversely, to pathologize moral wickedness is equally problematic. Then individual culpability becomes nothing more than mental illness requiring therapy, while corporate culpability becomes a social problem requiring policy changes.[54] Put differently, to suggest that all social evil is a consequence of natural evil is as problematic as suggesting that all natural evil is a consequence of social evil (e.g., God's punishment for sin).[55] To secularize natural evil, to regard it simply as part of nature, neither as God's punishment nor as the best possible world created by God, is to risk blurring the distinction between natural and social evil. If so, social evil may then be regarded as a more or less "natural," if not inevitable consequence of natural evil. This can only function as an excuse not to address anthropogenic climate change.

There can be no doubt that the Anthropocene in general and climate change in particular raise complex moral problems. Stephen Gardiner aptly describes this as a "perfect moral storm." He shows how three thorny "storms" intersect and exacerbate each other to create this perfect storm, namely: (1) the "basic storm," that is, the way in which causes (carbon emissions) and effects (changing climates) become spatially dispersed while agency becomes fragmented through a vast number of households, institutions, and countries; (2) the "inter-generational storm" (i.e., balancing current interests against the interests of future generations); and (3) the "ethical storm" in the sense that the dominant utilitarian framework (a cost-benefit analysis) to assess the situation is itself part of the problem. It cannot weigh current interests against long-term future interests or human interests against those of other forms of life. It also does not address the problem of moral corruption.[56]

Gardiner does not use the term "wicked problem" to capture what is at stake but it is at least clear why climate change poses both a moral and an ethical problem (in the sense of second-order ethical theories). To capture the complexity of the problem, he offers eight propositions that may be summarized as follows:

- Runaway carbon emissions are the most serious global environmental problem that humanity has ever faced.
- The framing of the problem in scientific and economic terms is dubious since the deepest challenge is ethical, namely, the need to protect the poor and vulnerable, future generations, and nonhuman forms of life.
- The problem poses a perfect moral storm given its global, intergenerational, and theoretical dimensions.
- The conceptual paradigm through which the moral problem is addressed, namely, a utilitarian cost-benefit calculus with reference to the well-known prisoner's dilemma and the notion of a "tragedy of commons" is inadequate because it does not address the intergenerational dimension of the problem.
- Ethicists reflecting on this "perfect moral storm" are not neutral observers but are themselves implicated, opening the door for self-serving analyses and ideological distortions.
- There is a real danger of considering "shadow solutions" that do not address the underlying problem. The failure of climate policies to curb carbon emissions and subsequent proposals for climate engineering may serve as examples.
- Because ethicists have become judges in their own case, ethical discourse on the perfect moral storm is in danger of adopting defensive strategies. It is therefore necessary to address bad arguments, forms of denial, and ideological distortions.
- The above does not imply that moral guidance cannot be offered to confront the challenges. The difficulty is that early guidance needs to be offered despite prevailing uncertainties.[57]

Again, Gardiner does not refer to the notion of wicked problems, but the eight propositions that he offers can clearly be added to the various lists of features of a wicked problem as discussed above.

What, then, is wicked about a wicked problem? My sense is that the term was originally used in a somewhat light-hearted, colloquial way, perhaps in jest, to refer to managerial problems that just cannot be fixed. Only a fool would take something all too serious, as it were not catching the joke. Amid a changing climate there is some need for comic relief. Jokes about what is devilish, demonic, wicked, or downright evil may well be the most appropriate genre to discuss such matters. Any internet search will reveal that such jokes abound.

If so, against the inclinations of the managerial "sciences" the term need not be used in any conceptually precise way as if it would be possible to state, quite exactly, what is wicked about a wicked problem and why climate change may be regarded as a wicked problem. The moral dimensions of the challenges posed by climate changes should be taken seriously, but perhaps there is no need to take the term "wicked problems" all that seriously. More precisely, I suggest that the typical characteristics of wicked and super-wicked problems are indeed helpful, while the term itself is less helpful than the many references to the literature seem to indicate. What should not be scoffed at is the managerial ethos that presumes that the challenges posed by climate change can be addressed so "exactly" through climate engineering or even social engineering. The danger is that the term "wicked problems" can lead to unnecessary confusion where a term such as structural violence or "a perfect moral storm" (Stephen Gardiner) would suffice to describe such problems. Except for a light-hearted moment, Occam's razor may need to be applied to avoid unnecessary additions to the vocabulary.

GLOBAL DIVIDES IN THE ANTHROPOCENE?

Third, one of the most significant debates on the Anthropocene is how such ruptures in Earth systems are shaped by social divides associated with race, ethnicity, class, gender, caste, culture, language, civilization, and religion.[58] For example: What is the gendered nature of the Anthropos who is able to alter Earth systems? How is an understanding of the Anthropocene constructed across global divides, for example, between the global North and the global South, developed and developing nations, industrialized and industrializing societies, those societies in the center of economic power and those on the periphery? How should the tendency in the global South to copy patterns of production and consumption in the global North be understood? Even if one may acknowledge global divides in terms of responsibility (read: culpability) for the root causes of the Anthropocene, such divides do not pertain to its consequences in the same way. To be sure those already vulnerable will suffer disproportionally while the elite may initially find ways to protect themselves from the worst. However, in the long run all humans, indeed all forms of life will be affected so that there is an unforeseen, terrible solidarity of "mutually assured destruction" between victims and perpetrators.[59]

As I noted above, it is clearly inappropriate to name an undifferentiated "anthropos" as the distinctive feature of the new geological epoch. At the same time, these forms of domination in the name of difference are clearly interrelated although at times also in tension with each other. But would it really help to pinpoint the underlying problem and name that accordingly, for example, as a White Anglo-Saxon Protestant problem? Such a focus would be too narrow and would not recognize the way in which one empire succeeded another throughout the Holocene epoch, the geological period within which human civilizations have emerged, thrived, and collapsed. Is the American dream worse than the *pax Romana*? What about the need to call the previously colonized to resist continuing colonial modes of thinking, as is evident in the writings of Franz Fanon and Steven Bantu Biko?[60]

Or is it, instead, a species problem? Is the problem human nature as such (being greedy, violent, and arrogant), applicable to all humans, or should the structural violence associated with the Anthropocene be regarded as a deviation from human nature so that such deviations may be regarded as culpable? If this is a deviation, how widespread is that deviation? How deeply is human nature rooted in mammalian, hominid, and hominin nature? If it is deeply rooted, is nature itself then the problem so that some human enhancement of nature is needed? Is this best regarded a fuzzy problem, where good intentions are mixed with evil ones and with inadvertent outcomes? Or can human behavior be understood as a function of insecurity and anxiety, the need to minimize pain and optimize happiness amid a struggle for survival?[61] If so, there is something tragic about the Anthropocene, but it would not do to locate the blame on some humans more than others. Then it is hardly a moral problem but one that requires, staying with medical terminology, a diagnosis and a remedy or therapy. Responsibility would then mean little more than being willing to go to the doctor and to take the prescribed medicine meticulously.

In raising these questions I admit that a fool can ask more questions in one paragraph than a wise person can address in a lifetime. It is nevertheless remarkable how frequently questions about human nature pop up in literature on the Anthropocene, in the humanities but also in the natural sciences. Indeed, what does it mean to be human in the age of humans? This is clearly not the kind of question that can be addressed in a paragraph or two, but for the sake of multidisciplinary collaboration on an appropriate diagnosis I do wish to offer two perspectives derived from Christian discourse on sin.

The one is to return to the distinction between sin as power and guilt (see the section on "Christian Vocabulary on Sin" in chapter 2). The structural violence associated with the Anthropocene is a mess[62] in which all humans, indeed all forms of life, find themselves stuck in, to which some have contributed more than others and under which some suffer more than others. Given the recognition of the role of Earth systems and its differentiated impact on particular ecosystems, it seems necessary to hold together a sense of the global and the local, how individual suffering may be a function of complex interactions.[63]

The other is to note the long-standing debates in Christian theology on the distinction between human nature and fallen human nature. This is arguably a debate between Manichean, Augustinian, Pelagian, and Irenean views over terms such as *posse peccare* (it is possible to sin; God's creation is "very good" but not "perfect"), *posse non peccare* (sin is contingent and therefore not necessary or inevitable), *non posse non peccare* (it is no longer possible not to sin), and *non posse peccare* (the hope that, in the end, it will no longer be possible to sin).[64] There seems to be as much confusion within discourse on the Anthropocene as there is in Christian theology over human agency! At least this should challenge any all too easy assumptions that humans are either self-interested, aggressive, and domineering or good-natured, cooperative, and compassionate by their very nature. It is also facetious to maintain that humans are a mixed bag of selfishness and kindness. There is a temptation to merely fluctuate between the pessimist assumption that people are generally so corrupt that power can never be entrusted to individuals, and the optimist assumption that the majority of people are ultimately good, just, and rational enough to come to appropriate decisions. As Slavoj Žižek observes, the democratic vision is based on both trust and distrust in human decision-making processes.[65] From a Christian perspective, the paradox is radicalized: humans are both fallen and, despite such fallenness, beloved by God.

To conclude, it should be clear that structural violence requires structural change. This is aptly expressed in the slogan "system change, not climate change." The diagnosis of structural violence requires a remedy at that level too. However, systems do not change by themselves. Change can be prompted by crises but an appropriate response to such a crisis is far from necessary. Structural change is not induced merely by policy making, by cultural trends, or by awareness raising through media campaigns, although these may each play a role. The underlying reason why structural change does not come easily may well have to do with reigning ideologies.

NOTES

1. Let me make this rather personal. According to my genealogical records I am a descendant of the following women in one way or another: Adriana Gabrielse van die Kaap (1662–), Angela van Bengale (–1720), Anna Groothenning van Bengale (1672–c. 1719), Armozyn Claasz van die Kaap (1661–c. 1733), Catharina van Malabar (1673–c. 1703), Catharina van die Kaap (1660–c. 1700), Catharina Vryman van die Kaap (–1700), Christina Susanna van die Kaap, Groote Catrijn van Palicatte (1631–1683), Helena van Malabar (1659–), Krotoa (Eva) van die Kaap (1643–1674), Lysbeth Vion van die Kaap (1698–), Maria van Batavia (c. 1665–c. 1717), and Regina Elizabeth Alhabama (c. 1735–). Krotoa was an indigenous Khoi woman. All the others were freed slaves or descended from slaves.

2. See Galtung, *Peace by Peaceful Means*, 31. See also his earlier essay "Violence, Peace and Peace Research," *Journal of Peace Research* 6:3 (1969): 167–91.

3. Ibid., 93.

4. See Dom Helder Camara, *Spiral of Violence* (London: Sheed and Ward, 1971).

5. Ibid., 55.

6. Paul J. Crutzen and Will Steffen, "How Long Have We Been in the Anthropocene Era?" *Climatic Change* 61 (2003): 251–57 (253).

7. Hamilton, *Defiant Earth*, 6.

8. See especially Mahmood Mamdani, "Reconciliation without Justice," *Southern African Review of Books* 46 (December 1996): 3–5; and many others who refer to his argument, recently also Allan Boesak in *Pharaohs on Both Sides of the Blood-Red Waters* (Eugene: Cascade Books, 2017), 117–46.

9. See the analysis in Ernst M. Conradie, *Reconciliation as a Guiding Vision for South Africa?*

10. Peralta and Mshana, *The Greed Line*.

11. See also Žižek, *Living in the End Times*, 132: "Consequently we should not say that capitalism is sustained by the selfish greed of individual capitalists, since that greed is itself subordinated to the impersonal striving of capital to reproduce and expand."

12. Konrad Raiser, "Towards Defining the Concept of Structural Greed," in *The Greed Line: Tool for a Just Economy*, edited by Athena Peralta and Rogate Mshana (Geneva: WCC, 2016), 91–93 (92). His list is quoted here verbatim.

13. Raiser, "Towards Defining the Concept of Structural Greed," 93.

14. Jung Mo Sung, "Toward a Conceptual Elaboration of Structural Greed," in *The Greed Line: Tool for a Just Economy*, edited by Athena Peralta and Rogate Mshana (Geneva: WCC, 2016), 94–99 (97).

15. Ibid., 97.

16. Ibid., 98.

17. See Yusoff, *A Billion Black Anthropocenes or None*, 6.

18. See Klein, *This Changes Everything*, 81.

19. Ibid., 89.

20. As quoted and discussed in Latour, *Facing Gaia*, 108.

21. See the executive summary of the Stern report at http://www.hm-treasury.gov.uk/sternreview_summary.htm (accessed June 7, 2019).

22. See Klein, *This Changes Everything*, 148.

23. Jared Diamond, *Collapse: How Societies Choose to Fail or Survive* (New York: Penguin Books, 2005).

24. See Wood, holding both views in *Deep Time, Dark Times*, 18, 45.

25. See, in the context of ecotheology, for example, the provisional comments by Jenkins, *The Future of Ethics*, 149–89; Conradie, *Redeeming Sin?*, xvi–xix; O'Brien, *The Violence of Climate Change*, 17–39; and Carol Wayne White, "The Wicked Problem of Climate Change: Re-Envisioning Hope: Anthropogenic Climate Change, Learned Ignorance, and Religious Naturalism," *Zygon* 53:2 (2018): 570–84. In each of these contributions wicked is assumed to mean "serious," while moral connotations are not avoided. This stands in contrast with the literature discussed below. A version

of this discussion on wicked problems will be published in a special edition of the journal *Philosophia Reformata* under the title, "Why, exactly, May Climate Change be Regarded as a 'Wicked Problem'?"

26. See C. West Churchman, "Wicked Problems," *Management Science* 14:4 (1967): 141–42, doi:10.1287/mnsc.14.4.B141.

27. This is how the Wikipedia entry has it. See https://en.wikipedia.org/wiki/Wicked_problem (accessed December 13, 2018). This quite comprehensive entry is indeed helpful as a starting point to the early literature. I also acknowledge Clive Pearson and Sytse Strijbos who guided me to some of the available literature.

28. Churchman, "Wicked Problems," 142.

29. See Horst Rittel and Melvin Webber, "Dilemmas in a General Theory of Planning," *Policy Sciences* 4 (1973): 155–69.

30. In his book *A Guide for the Perplexed* Ernst F. Schumacher distinguishes between divergent and convergent problems. Convergent problems are those for which attempted solutions gradually converge on one solution or answer (e.g., the mechanics of a bicycle). Divergent problems are those which do not converge on a single solution; different answers appear to increasingly contradict each other as they are further developed (e.g., models of education). For Schumacher, convergent problems are related to the nonliving world, while divergent problems are concerned with living entities. This correlates with his views on four levels of being, namely, the mineral sphere (ruled by cause and effect), the plant sphere (rules by stimulus), the sphere (ruled by motives), and the human sphere (shaped by will), suggesting a movement from passivity to activity. See Ernst F. Schumacher, *A Guide for the Perplexed* (New York: Harper & Row, 1977), 121–37.

31. Rittel and Webber, "Dilemmas in a General Theory of Planning," 160.

32. Ibid.

33. Ibid., 161–69.

34. Ibid., 160.

35. See Russell L. Ackoff, *Redesigning the Future* (London: Wiley, 1974), 21–22.

36. Ibid., 30.

37. Ibid., 32.

38. Ibid.

39. As an aside, David Wood notes, "The promise of Reason may well have been hijacked by its bastard son Calculation." See Wood, *Deep Time, Dark Times*, 28.

40. For one example, alongside many others, see Morton, *Dark Ecology*, 36–37.

41. See the undated, archived publication of the Australian Public Service Commission entitled "Tackling Wicked Problems: A Public Policy Perspective" at https://www.apsc.gov.au/tackling-wicked-problems-public-policy-perspective (accessed December 13, 2018).

42. See Richard James Lazarus, "Super Wicked Problems and Climate Change: Restraining the Present to Liberate the Future," *Cornell Law Review* 94:5 (2009): 1153–233 (1160).

43. Ibid., 1160.

44. Ibid., 1161.

45. See also Morton, *Dark Ecology*, 37.

46. See Kelly Levin, Benjamin Cashore, Steven Bernstein, and Graeme Auld, "Overcoming the Tragedy of Super Wicked Problems: Constraining our Future Selves to Ameliorate Global Climate Change," *Policy Sciences* 45:2 (2012): 123–52 (123).

47. Quoted in Willis Jenkins, Evan Berry, and Luke Beck Kreider, "Religion and Climate Change," *Annual* Review *of Environment and Resources* 43 (2018): 9.1–9.24. The reference is to Mike Hulme, "Climate Change: Varieties of Religious Engagement," in *Routledge Handbook of Religion and Ecology*, edited by Willis Jenkins, Mary Evelyn Tucker, and John Grim (New York: Routledge, 2017), 239–48 (246).

48. Žižek, *Living in the End Times*, 5.

49. See Mary Midgley, *Wickedness: A Philosophical Essay* (New York: Routledge, 2001), vii.

50. Ibid., 74–94.

51. Ibid., 194, 204.

52. Jenkins, *The Future of Ethics*, 21–22.

53. See Neiman, *Evil in Modern Thought*, 287.

54. Midgley, *Wickedness*, 51.

55. See Neiman, *Evil in Modern Thought*, 23.

56. See Gardiner, *A Perfect Moral Storm*, 19–48.

57. See Ibid., xi–xiv.

58. Galtung develops his theory of conflict in relation to numerous other themes, including peace, democracy, human rights, development, and civilization. It outlines the forms of domination associated with contrasting civilizations. He distinguishes between Occident (Greco-Roman, influenced by the Abrahamic religions), Indic, Sinic, Buddhic, and Nipponic types of civilization, admitting to leaving out of the picture various indigenous forms of civilization, including African civilizations). See the fourth part of Galtung's, *Peace by Peaceful Means* (1996), 196–264.

59. See Latour, *Facing Gaia*, 216.

60. See Steven Bantu Biko, *I Write What I Like: Selected Writings* (Johannesburg: Picardo Africa, 2017); Franz Fanon, *The Wretched of the Earth*, translated by Richard Philcox (Cape Town: Kwela Books, 2017).

61. Accordingly, as Michael Taylor seems to suggest, greed and self-centered behavior might be better understood as fostered and provoked by insecurity than as a form of moral perversity. If so, our moral duty is not to change human nature but to imagine and implement systems and structures that at best encourage us to move away from greed toward justice, sustainability, and well-being. See Michael Taylor, "On Greed: Toward 'Concrete and Contemporary Guidance for Christians,'" in *The Greed Line: Tool for a Just Economy*, edited by Athena Peralta and Rogate Mshana (Geneva: WCC, 2016), 80–90 (88).

62. On the notion of a "mess" (a system of interrelated problems), see again Ackoff, *Redesigning the Future*, 21.

63. I have adopted and adapted such a notion of sin as power and as guilt in all my recent work on "redeeming sin." See especially *Redeeming Sin?* (2017).

64. See especially chapter 3 (on the different storylines) and chapter 5 (entitled "Posse non peccare?") of *Redeeming Sin?*

65. Žižek, *Living in the End Times*, 5.

Chapter 5

Distortive Ideologies in the Anthropocene?

ON IDEOLOGY IN THE ANTHROPOCENE

There is a long list of ideologies prevalent in the twentieth century that remain candidates to describe the distortion of human consciousness in the Anthropocene. Capitalism, imperialism, colonialism, sexism, racism (white supremacy), fascism, communism, various stripes of nationalism,[1] state security, heteronormativity, consumerism, name it! Note that "ideology" (literally the study/logic of ideas) is used here in the pejorative sense and not in the more neutral sense of the history of ideas or more generally views of the world (*Weltanschauungen*). It is a matter of meaning in the service of power.[2] However, two questions have to be raised with regard to the ideologies listed above: First, to what extent do these shape human behavior subconsciously so that the distortions that they bring are not widely acknowledged? If these ideologies are well known, then they may still express, condone, or legitimize interests and power but are no longer operating subconsciously. Conscientization then cannot be the key to overcome the problem.[3] Second, to what extent are these the ideological forces that have introduced what is new to the Anthropocene? Would industrialized capitalism suffice to explain what has gone wrong? What changed with the advent of the Anthropocene?

It may be helpful to draw again on the work of Johan Galtung, widely regarded as the founder of peace studies. In *Peace by Peaceful Means* (1996) he offers an analysis of a wide range of ideologies and places them in the context of human evolution. His analysis is worth quoting at some length:

> We human beings are parts of nature, of the "earth" from which we have come and to which we return. How can this human-nature holon live in harmony,

> without direct and structural violence? As small bands of human gatherers (not hunters who exercise direct violence against fellow life) in a sufficiently generous nature context? Our present reality is far from that. Rather, analytically, we will need to think in terms of at least ten fault-lines criss-crossing that human/nature holon: humans/non-humans; gender (male/female); generation (old/young); race (white/colored); class (high/low); nation (high/low); countries (center/periphery); and the three edges in the state-civil society-capital triangle, in social and world space.[4]

Galtung then correlates these fault-lines with the ideologies of speciesism, sexism (patriarchy), ageism (gerontocracy), racism, classism, nationalism, territorialism, statism, anarchism, capitalism, and super-statism (imperialism), each carrying possibilities for direct, structural, and cultural violence. He adds that "The order mirrors the evolution of the human condition: first nonhumans (abiota and biota, divided into micro-organisms, plants, and animals); then humans of two genders, reproducing in generations; possibly emerging as different races in different places; then class differentiation as nomads who become pastoral nomads who then become pastoral and sedentary, attached to territory with borders; then the state as territorial locus of power within a territory over non-state, and as super-states over states."[5] Galtung continues to discuss various forms of state systems—ranging from dissociative to associative, confederal, federal, unitary states, and super-states. He observes that both the peace-building capacity and the peace-threatening capacity of such state systems increase with the degree of integration and that there is "no reason to believe that super-states are any more peaceful than states."[6] The formation of super-states merely transforms the scale of possible conflict.

This suggests that globalization may well be a necessary component in the transition from a species that not only affects local and bioregional ecologies but that has an impact on Earth systems. As Slavoj Žižek remarks, the fundamental lesson of globalization is that capitalism can accommodate itself to all civilizations, from Christian to Hindu or Buddhist.[7] He speaks of its "infinite adaptability" including an ecological reorganization of capitalism and warns against attempts to de-ideologize climate change by treating it as a technical issue.[8] Again the question needs to be raised as to what ideologies arise with the advent of the Anthropocene that go beyond the conditions of the late Pleistocene and Holocene? This question should remain open to contestation, but two candidates may be provisionally identified, namely, the assumption that unlimited economic growth is possible and necessary and the assumption that life is epitomized by human life, that is, humanism.

ON ECONOMIC GROWTH

The first candidate for an ideology that speaks directly to the Anthropocene, insofar as this is tied to the Great Acceleration, is the assumption that sustained economic growth is a necessary ingredient of a flourishing economy and that this is not only necessary to address poverty, unemployment, and inequality but also to generate the capital that will be needed to transform the energy basis of an industrialized economy, and to help address the ecological side effects of such an economy. Most modern economies seek to address unemployment by promoting higher economic growth, but this hardly resolves the problem, paradoxically at a time that a greater workforce is needed for care-giving and environmental restoration.[9] In a survey of the history of environmentalism since 1946 John McNeill and Peter Engelke conclude that "the global economy continues to expand in ways that threaten all that environmentalists cherish." They add, "The postwar vision on unending economic growth and unbridled technological progress remains intact—if no longer unchallenged."[10]

Following the discussion on shifts in discourse on sustainability in the previous chapter it is now necessary to clarify what limits to economic growth are at stake and why the failure to recognize such limits may be described as an ideology. An early indication of the intuition that a finite planet cannot sustain expanding demands on its resources is expressed in a quite lucid way in a story about Gandhi:

> After Gandhi led India to independence someone asked him whether India will now try to reach the same standard of living as England. His answer was: "It took Britain half the resources of the planet to achieve its prosperity; how many planets will a country like India require?"[11]

The original report on *Limits to Growth* focused on nonrenewable resources that are put under further pressure by population growth. Subsequent debates shifted to the sustainable use of renewable resources (agricultural land, water, fisheries, forests) and the call for sustainable development. This is widely regarded as a key concept to redirect the global economic order toward an alternative that is more sustainable in terms of the Earth's resources. It gained currency in a report from the United Nations' Commission of Environment and Development (chaired by Gro Harlem Brundtland) entitled *Our Common future* (1987). It was especially prominent in the theme of the World Summit on Sustainable Development held in Johannesburg in 2002, guided the UN's Millennium Development Goals (2000) and remain the focus of the Sustainable Development Goals (2015).[12]

The value of the notion of sustainable development is that it may serve as a corrective against expansionist notions of economic growth that disregard the environmental impact of such economic activities. Nevertheless, such a notion of sustainable development typically assumes the need for sustained economic growth to allow for such "development." The Brundtland report asserted that growth has "no set limits in terms of population or resource use beyond which lies ecological disaster."[13] Some critics therefore regard this as an attempt at the greening of global capitalism, as a euphemism used by entrepreneurs for "business as usual," namely, an emphasis on economic growth, qualified by a few environmental cautions. When faced with a choice between development and a sustainable environment, the interests of developers and entrepreneurs (who often provide short-term economic gain in terms of employment) regularly seem to receive a priority. The underlying development paradigm, with its strong emphasis on economic growth and market expansion, serves primarily the interests of powerful corporate-driven, market-oriented economic forces. Such critics maintain that there is a tension between economic growth and ecological sustainability and raise the question whether sustainable development, insofar as it is based on economic growth, is not a contradiction in terms,[14] maintaining two incompatible goals, namely, to keep developing and to leave the biophysical environment more or less intact. They argue that it cannot address social inequalities or the root causes of environmental degradation. The notion of sustainable development does not escape from conceptual contradictions that would apply to prospects for "sustainable growth."[15]

Given its track record and despite huge investments in finance, staff, and infrastructure, the "failure of development" (to perpetuate instead of reduce inequalities) cannot be rectified by making that failure more sustainable![16] The underlying problem is that the dominant paradigm of development is premised on Western notions of progress and linear ideas of history.[17] Development assumes the need for economic growth and economic growth assumes the possibility of scientific and technological progress. Accordingly, if such progress can be sustained, growth and therefore development can be sustained. To suggest that there are limits to economic growth therefore seems to impose limits on progress. As Carmen Gonzalez observes,

> The root cause of the contemporary ecological crisis is an international economic order premised on unlimited economic growth that impoverishes the global South and facilitates the overconsumption of the planet's resources by its more affluent inhabitants. This economic order reinforces the colonial notion that all societies must evolve through particular stages until they achieve the apex of civilization represented by the global North. It casts development as the

ubiquitous goal of all states and peoples, and equates development with rising material consumption.[18]

Instead, one may argue that the sustainability of natural ecosystems should be regarded as the point of departure.[19] In this vein, Leonardo Boff concludes that sustainable development is an oxymoron that leads to confusion; it does not symbolize a new way of looking at the world.[20] It is little more than a way of retarding the collapse of modernity.[21] Christophe Bonneuil and Jean-Baptiste Fressoz put it bluntly: "The Anthropocene thus cancels the peaceful and reassuring project of sustainable development," built as it is on maximal sustainable yield.[22] Perhaps the underlying problem is that "sustainable development" does not escape from the ambiguity of the notion of "development" itself. It seems to assume that the meaning of "development" is clear enough and that such development now has to become more sustainable.[23] The assumption is still that industrialized economies may serve as the norm for "developing" countries.

It is not necessary to explore the ambiguities embedded in the notion of development in the context of discourse on the Anthropocene in any detail here. The debate on whether sustained economic growth is possible is situated within the tension between two contrasting insights. On the one hand, there is, in general, indeed a strong correlation between economic growth, the use of energy and environmental impact, most notably carbon emissions. On the other hand, this is not a direct correlation. The price of nonrenewal resources remains stable relative to other inputs such as technology, capital, and labor. Forms of economic growth based on the knowledge economy have limited environmental impact. Moreover, wealthy countries seem to be able to address immediate environmental problems more successfully. Lawrence Summers, the former chief economist of the World Bank, even suggests that "there are no . . . limits to the carrying capacity of the Earth that are likely to bind at any time in the foreseeable future. . . . The idea that we should put limits on growth because of some natural limit is a profound error."[24] According to this (rather racist) logic, a lack of economic growth is related to a lack of education and technological know-how and ingenuity, perhaps an insufficient work ethic, or incompetence, maladministration, and corruption. Likewise, the assumption that there are limits to economic growth in terms of resources is questioned in *An Ecomodernist Manifesto*: "To the degree to which there are fixed physical boundaries to human consumption, they are so theoretical as to be functionally irrelevant."[25] On this basis, the *Ecomodernist Manifesto* expresses the belief that "both human prosperity and an ecologically vibrant planet are not only possible but also inseparable." Accordingly, it is possible to "decouple human well-being from environmental destruction" and therefore to "embrace an optimistic view toward human capacities and the future."[26]

In the somewhat more sober assessment of Jeffrey Sachs, economic growth is indeed required for economic development. The key driver for shared prosperity is technology transfer because such technology allows an economy to derive more value at a given level of inputs. He adds: "The fundamental reason for believing that prosperity can spread to all corners of the world is that the very science and technology that underpin prosperity in the rich world are potentially available to the rest of the world as well. If rich countries are rich because they have adopted these technologies—power generation, medicine, transport, construction and much more—the same improved technologies can also be adopted in today's impoverished countries."[27] He resists the "failed Marxist notion" that the rich are rich because they exploit the poor (in terms of resources, labor, and markets) and instead mentions, besides population pressures, the role of savings and investment, exports and imports, public capital and adaptation to local geography as the most significant "hurdles" for "climbing the development ladder" (with the United States at the top!).[28] Is technology then the source of "salvation," understood as economic growth, that will ensure "development"? Or is the "inadvertent" use of technology precisely the root of the underlying problem?

By contrast, in a document entitled *Climate Change – A Challenge to the Churches in South Africa* produced by the South African Council of Churches (Marshalltown: SACC, 2009) the following ten factors that contribute to the production of wealth are identified:

- access to energy sources (consider the availability and use of fossil fuels);
- natural resources, including various nonrenewable resources and renewable resources such as water, soil, trees and plants, and fish stocks (access to cheap sources was a driving force behind colonialism, remains crucial for many industries, leads to a current scramble for raw materials available in various African countries and are often monopolized by powerful economic institutions);
- land/property (consider the conquest of land through war and imperialism, ensuring political control over such resources);
- labor and employment (consider the role of slavery, the exploitation of cheap labor, the tendency toward outsourcing labor costs, the impact of trade unions and the cutting of labor costs by replacing it with technology);
- knowledge and skills (consider the role of research, ingenuity, education and training, and unequal access to quality education);
- the means of production (including various forms of technology that are sometimes beneficial, sometimes destructive);

- the formation of strong institutions such as business corporations with efficient management systems which allow for the sharing of risks and give some groups a competitive advantage over others (consider the ways in which the globalized economy is dominated by powerful multinational companies and the interests of their shareholders);
- cultural values and virtues (consider the emphasis on entrepreneurship, creativity, innovation, dedication, diligence, productivity, efficiency, innovative leadership) as well as economic drives (consider the quest for progress, success, fame, affluence);
- the availability of capital to finance large new projects based on previous profits (consider the role of credit, judgments on creditworthiness, and interests paid on such credit) as well as the role played by investments, shareholding and the quest for profit, interests on investments, and the lucrative transfer of money or financial products where no trade in goods is involved (i.e., making money from money alone, without adding anything to economic well-being);
- access to viable markets to sell products (influenced by legislation and trade agreements) and the stimulation of consumer demands through advertising.[29]

The document emphasizes that the production of wealth always depends on an interplay between all these factors. Each of these factors is deeply influenced by historic injustices and the legacy of imperialism, colonialism, classism, racism, sexism, and cultural elitism. In a telling comment the document adds that "some may wish to emphasise the role of cultural values, virtues and the proper management of corporate institutions. Thereby they tacitly imply that those who are poor are relatively lazy, stupid, slow, corrupt or unlucky. By contrast, those of us who emphasise injustices typically focus on all the other factors, including access to the means of production, labour costs and unfair trade relations."[30] Debates on the interplay between these factors remain unresolved.

Where, then, do the limits to growth really lie, if not with the availability of nonrenewable resources? One option is to focus on the history of colonization, the exploitation of "cheap" resources through military conquest, and the exploitation of cheap labor. If so, the Great Acceleration corresponds to the capturing of ecological surpluses of low-income countries by Western industrialized countries.[31] Accordingly, economic growth is only possible at the cost of economic inequality. Similarly, one may focus on market regulations and control by powerful nations and corporations: economic growth is only possible through the availability of new markets for the distribution of consumer goods. Another option is to focus on the failure to include the true costs for "what nature provides" in economic reasoning. This would include all the four elements of nature, that is, air (climate change), fire (energy), water (rivers, oceans, underground water), and land (food, biodiversity). Alternatively, one may argue that the problem does not lie with "natural resources" but with strained "ecosystem services." One may therefore focus on the limited capacity of the biosphere to absorb the waste products of the global economy, including carbon dioxide.

In my view, it may help to focus on what is understood under economic growth itself. Part of the problem is that such growth is typically still measured in terms of GDP which does not differentiate between "products" that are harmful or beneficial to (human) well-being. It is therefore unable to indicate whether a form of growth is desirable or not. It does not take voluntary work into account and the utilitarian calculus employed cannot measure the many things that do contribute to human well-being, including stable relationships, friendships, satisfying work, loyalty to institutions, opportunities to express one's talents and dedication to some higher cause. There is ample room for growth in human well-being. This can be either served or hampered by economic activity, technological innovation, and cultural practices.

The most pertinent question is whether and to what extent economic growth assumes growth in biophysical throughput. The clearest measure of that remains the use of fossil fuels as sources of energy. It is indeed possible to decouple economic growth from the use of fossil fuels. This can be done in several ways, for example, by (a) increasing the efficiency of resource use through technical progress, (b) substituting damaging inputs for less damaging inputs (e.g., using renewable energy sources instead of fossil fuels), and (c) changing the composition of economic output from goods to services to intellectual content (requiring less raw materials).[32] The last of these requires a process of dematerialization and more specifically decarbonization in economic production.[33] This requires as a first step carbon density, by using less fossil fuels per unit of production. Put together, a rapid decoupling[34] of economic growth and environmental impact is indeed technically feasible and economically possible, if not politically likely. As Johan Rockström and colleagues observe, "Although the Paris Agreement's goals are aligned with science and can, in principle, be technically and economically achieved, alarming inconsistencies remain between science-based targets and national commitments."[35] The ecomodernist optimism that such rapid dematerialization is feasible reflects a knowledge-based economy and the cultural patterns of the West and the East rather than the South. The impact of such a knowledge-based economy on unemployment and inequality poses almost unassailable challenges.

Insofar as these steps can be implemented, it may be possible to sustain economic growth, especially through a knowledge-based economy in already highly industrialized countries. Mark Swilling and Eve Annecke explains what this would require:

> If the transition to a sustainable socio-economic regime means anything, it will mean investing in innovations that will demonstrate practical ways of promoting non material economic growth, which will require minimal quantities of non-renewable resources inputs, the reuse of all waste outputs, and replenishment of renewable resources that have been over-exploited (such as fish, soils, water supplies, air quality, biomass, and biodiversity.[36]

However, the reality is that such a decoupling of economic growth and biophysical throughput is not yet evident at a global scale.[37] In industrializing countries (including China and India), economic growth remains coupled with biophysical throughput. Neither a postagricultural nor a postindustrial society is possible, even though it is true that the share of these two sectors in GDP has been shrinking in the context of the third and fourth industrial revolutions. The best yardstick of the coupling of economic growth and biophysical throughput remains the use of fossil fuels, as may be measured by annual carbon emissions and the concentration of carbon dioxide and other greenhouse gases in the Earth's atmosphere. Here is the latest verdict produced by the Global Carbon Project in this regard:

> Global CO_2 emissions from fossil fuels and industry have increased every decade from an average of 11.4 $GtCO_2$ in the 1960s to an average of 34.7±2 yr^{-1} during 2009–2018. Emissions in 2018 reached a new record high of 36.6±2 $GtCO_2$ with a share of coal (40%), oil (34%), gas (20%), cement (4%), and flaring (1%). Global emissions in 2019 are projected to increase by an additional 0.6% (−0.2% to +1.5%), a slower growth than in the past two years.[38]

Given this correlation between economic growth and the use of fossil fuels, one has to wonder whether the assumption that economic growth is necessary for (human) well-being is not the major stumbling block for addressing one of the most obvious symptoms associated with the Anthropocene, namely, the growth of greenhouse gases in the Earth's atmosphere. To put this proverbially: it is as if a smoker, alcoholic, or drug addict complains about the symptoms associated with their addictions but focus on the possibility that some of the symptoms may be unrelated to the frequency of smoking, drinking, and the use of drugs. There is an obvious need to confront the main cause of the problem as a function of population, production, consumption, and technological power.

Let me illustrate this danger with reference to Mark Lynas's argument in *The God Species* which adopts a tone of technological optimism that departs from his earlier books. Lynas offers a popularized account of the concept of planetary boundaries. He describes these boundaries in some detail with reference to the work of Johan Rockström, Will Steffen, and others. He compares that to the 1972 *Limits to Growth* report and insists that the recognition of planetary boundaries does not necessarily imply any limit to economic growth or productivity. Instead, it circumscribes a safe space within which the economy can be structured and within which humans can flourish indefinitely. Planetary boundaries need not constrain humanity's potential or ambition. He adds this: "Nor does it necessarily mean ditching capitalism, the profit principle, or the market, as many of today's campaigners demand. Above all, this is no time for pessimism: We have some very powerful tools available to allow us to live more gently on this planet, if only we choose to use them."[39] He speaks approvingly of economic growth and technological innovation as ways to solve pressing environmental challenges. I will return to the use of such tools below, but his "pro-growth perspective" is pertinent here.

In the conclusion to his book, Lynas offers three arguments for this. First, growth in consumption, like growth in population, tends to taper off once a certain level is achieved. Arguments against overconsumption only make sense in affluent countries and are restricted to that. Second, reducing poverty requires raising levels of consumption. The assumption that people in less affluent countries will limit their aspirations for prosperity by emulating the example of those who deliberately consume less is fanciful and patronizing. Opposing growth in developing countries is a nonstarter. Instead, the rising consumption of energy should be channeled toward low-carbon sources. Third, there are no viable alternatives to the emphasis on economic growth. The only alternative to the modern market system is painful contraction, unemployment, and political instability, as illustrated by the major recessions since 1930. Growth is therefore central to successful capitalist economies, although Lynas regards the trend toward dematerialization (becoming less resource intensive per unit of output) as positive for several planetary boundaries. Technology will help here, for example, in the form of music downloads (instead of CDs), online newspapers and e-books (instead of printed versions), and virtual meetings and conferences. He admits that energy needs will rise to make a prosperous lifestyle for all humans possible, but promotes the use of nuclear (fission) and solar (fusion) energy as alternatives to fossil fuels over the next three decades.[40] On this basis, he remains confident that human aspirations need not be constrained and that economic growth can be compatible with respect for planetary boundaries.[41] He offers the following conclusion:

> The planetary boundaries provide a physical and ecological limit to how far humans can trespass on the biosphere, but if we respect them fully then I believe that growth—as currently conceived—can continue more or less indefinitely. In

> contrast to many environmentalists, I do not see any convincing ecological reason why everyone in the world should not be able to enjoy rich-country levels of prosperity over the half-century to come. None of the planetary boundaries rule out this leap forward in human development—and as far as the Earth system is concerned, it is the boundaries that must provide the ultimate guidelines to the human project. . . . We can respect all the boundaries at the same time as eliminating poverty, allowing for a peak in world population around mid-century and within the context of an economic system that requires growth.[42]

Let me agree with Lynas, for the sake of the argument, that economic growth is compatible with planetary boundaries on condition that such growth is decoupled from growth in biophysical throughput. Let me then ask Lynas this question: If one is confronted with a choice between allowing for rising carbon emissions for the sake of growing prosperity (albeit only for some) and of limiting economic growth for the sake of staying within planetary boundaries, what would you choose? I presume he would insist that there is no need to choose in this regard. However, at this stage, with carbon emissions still rising annually, this remains a theoretical possibility only. The reality is that this is a choice being made annually at every meeting of the Conference of the Parties in terms of the UN Framework Convention on Climate Change, in the hope that such a theoretical possibility will come to fruition before too many of the planetary boundaries are breached.

This leads me to a further question. Is the failure, indeed the refusal, to confront the assumption that sustained economic growth is necessary for human well-being not indicative of an ideological distortion, and not merely of a technical, economic, or political problem? The recognition of ideological distortions regarding economic growth may be illustrated with a few pointers to the available literature where this is expressed:

As a first witness to the stand, Johan Galtung argues that the emphasis on economic growth is deeply embedded in the cosmology of what he calls the Occident, that is, the world of modernity, influenced by its Greco-Roman heritage and all the three Abrahamic religions. He argues that continuous economic growth is embedded in a particular understanding of time as *Chronos* and as *Kairos*:

> In the construction of Time there is an Idea of Progress, at both the Personal and the Social levels, building strong persons and strong societies, under an impending Crisis confronting Occident I with "make it or break it." Strong people who have made it become social Elites, and strong societies that have made it come to the World Center. In the process Nature is subdued, also by an anti-holistic and anti-dialectic Episteme in the service of development as nature, personal, social, and world engineering. Guiding development is one, singular,

> universally valid Principle: to realize the Rule of God, personally and socially in the religious version (with Clergy on top), and to realize the Rule of Growth (with Money on top), personally and socially, in the secular version. The social and world Periphery consists of those people and societies that accept the guiding Principle; others, who do not accept the Principle, are Evil. Peace implies their elimination / incapacitation / marginalization, so that development can proceed unimpeded by evil forces. And this may have catastrophic implications for Other.[43]

Galtung adds that in this context the need for economic growth, labor productivity, and the quest for material prosperity (manifested in a consumerist culture) function as godlike principles. The quest for material prosperity indeed functions as the secular successor to striving in one's daily work to the glory of God: "individual well-being in the sense of a high standard of individual material life, comfort; and if not eternal afterlife, at least longevity and good health. Material standards of living play very much the same role for development as freedom in connection with the pursuit of peace: the overriding goal that justifies the negative side-effects, such as eco-crisis, misery, repression, and war."[44]

A second witness is the Dutch economist Bob Goudzwaard. In his early book on *Idols of our Time*, he explores the ideologies of revolution, nationalism, material prosperity, and military security. With regard to prosperity he observes that national economies may accept moderation if need be, regard a successive decline in the size of the economy as a recession, and would never promote a shrinking economy. He argues that "we" (he focuses here on Westerners) have put our faith in rising economic growth and technological progress. This ideological commitment disables an "economy of enough," that is, one based on Torah notions of care, gratitude, satisfaction, and sufficiency.[45] While economic growth supposedly creates happiness, it is unhappiness with what one has that now sustains economic growth. He adds this:

> Our current economic thinking and social system works in the opposite direction of the Torah. Maintaining economic growth is our first priority. Then arises questions of what to do with the resulting unemployment, the dehumanization of work, the rape of the environment and other problems. Our priorities constrain us to use increasingly large-scale technologies ad to accept weapons contracts from wherever they come. These are obvious signs of our slavery to an ideology of prosperity.[46]

A third important witness is an authoritative article on the advent of the Anthropocene by Will Steffen, Jacques Grinevald, Paul Crutzen, and John McNeill. In a remarkable conclusion they observe that the concept of the

Anthropocene directly challenges neoclassical economic theory. They even liken this trust in notions of progress to climate skepticism and to the challenges posed by Darwinian theory to (Christian) understandings of human origins:

> Climate scepticism, or more appropriately the denial of contemporary climate change and/or its human causes, is, in many cases, a classic example of "cognitive dissonance"; that is, when facts that challenge a deeply held belief are presented, the believer clings even more strongly to his or her beliefs and may begin to proselytize fervently to others despite the mounting evidence that contradicts the belief. This response may become even more pronounced for the Anthropocene, when the notion of human "progress" or the place of humanity in the natural world is directly challenged. In fact, the belief systems and assumptions that underpin neo-classical economic thinking, which in turn has been a major driver of the Great Acceleration, are directly challenged by the concept of the Anthropocene. . . . The ultimate drivers of the Anthropocene . . . if they continue unabated through this century, may well threaten the viability of contemporary civilization and perhaps even the future existence of Homo sapiens.[47]

A fourth witness is the sobering critique of the growth fetishism offered by Clive Hamilton in *Requiem for a Species*. He observes that an object becomes a fetish when magical powers are ascribed to it. He then argues that this applies to economic growth, especially as expressed in GDP growth. This is even clad in religious terms so that such growth may have quasi-soteriological significance (see the next chapter). The need for such growth expresses a deeply held assumption that the Earth's resources are well-nigh infinite and that humans have the right to exploit them without restraint for their benefit.[48] Note that such growth is regarded not as the cause of environmental destruction but precisely as the only available solution to it. Hamilton comments:

> In affluent societies religious value seems to be invested in the most profane object, growth of the economy, which at the individual level takes the form of the accumulation of material goods. Our political leaders and commentators believe that it has magical powers that provide the answer to every problem. Growth alone will save the poor. If inequality causes concern, a rising tide lifts all boats. Growth will solve unemployment. If we want better schools and more hospitals then economic growth will provide. And if the environment is in decline, then higher growth will generate the means to fix it. Whatever the social problem, the answer is always more growth.[49]

After surveying various financial models to address mitigation targets and adaptation costs Hamilton concludes that the main obstacle to implementing climate policies is not that it may curtail economic growth, whether measured in terms of GDP or not. The problem, he argues, is the fixation with economic growth, the growth fetish, "the unreasoning obsession that arises because growth is believed to have magical powers"—to improve living standards, to fulfill dreams of prosperity, and even to confer cultural superiority.[50] If so, growth has indeed become ideological, a form of totemism, to the point that the need for such growth is taken for granted without further reflection on the correlation between GDP growth and human well-being. In *Earthmasters* Hamilton employs a secularized version of the Christian vocabulary of sin-talk to make the same point:

> To question the possibility of, let alone the need for, endless expansion is to commit the modern sin. Those who seem to challenge this most holy of truths must be cast out. This is the only explanation I can think of for the extraordinary hostility directed by some at environmentalists . . . [who] have become the scapegoats for our sins, loaded up with our own guilt and sent off to the wilderness.[51]

Yuval Harari, another witness to the stand, concurs. The phenomenal economic growth of the last century or so has enabled humanity to produce abundant food, medicine, energy, and gadgets to make life easier (at least for some). This may be the reason why, as Harari wryly observes, when politicians, CEOs, and voters have to choose between economic growth and ecological stability, they almost always choose growth.[52] In fact, Harari argues that

> Much of the credit for overcoming famine and plague belongs to the ardent capitalist faith in growth . . . capitalism did make an important contribution to global harmony by encouraging people to stop viewing the economy as a zero-sum game, in which your profit is my loss, and instead see it as a win-win situation, in which your profit is also my profit. This mutual-benefit approach has probably helped global harmony far more than centuries of Christian preaching about loving your neighbour and turning the other cheek.[53]

In *This Changes Everything* Naomi Klein observes that the global economic system (unregulated capitalism) is at war with the planetary system. It would not help to rebrand or green capitalism. The profit motive cannot serve as the midwife for the transformation of energy basis of the global economy that is required.[54] She explains:

> Our economic system and our planetary system are now at war. . . . What the climate needs to avoid collapse is a contraction in humanity's use of resources; what our economic model demands to avoid collapse is unfettered expansion.

> Only one of these two sets of rules can be changed, and it's not the laws of nature. . . . The problem, however, is that this scale of economic planning and management is entirely outside the boundaries of our reigning ideology. The only kind of contraction our current system can manage is a brutal crash, in which the most vulnerable will suffer most of all.[55]

Yet another witness is the ecosocialist author, editor, and activist Ian Angus. He offers a review of literature on economic growth and the environmental impact of the extraction of resources and the processes of production, distribution, and consumption. He notes that two explanations for the assumption that the need for economic growth is nonnegotiable may be offered. Some refer to human nature to argue why economic growth is capitalism's way of meeting human desires because "enough is never enough."[56] Others argue that this assumption is a fundamental mistake, that we have been seduced by a false ideology. The drive toward growth is a fetish, a psychological obsession, an addiction, even a spell. Angus observes that under conditions of capitalism economic production is no longer aimed at consumption, but at exchange. Capital is used to exploit labor and nature in order to produce goods that can be sold for more than the costs of production. This allows for an accumulation of capital that repeats itself endlessly. On this basis, Angus concludes that the ideology that economic growth is necessary for prosperity does not *cause* perpetual accumulation; it *justifies* it.[57] Likewise, one may say that the ideology of apartheid did not cause domination in terms of race and ethnicity. It served as a justification of it and in this way reinforced the structural violence associated with that dispensation.

In their book *The Human Planet: How We Created the Anthropocene* (2018) Simon Lewis and Mark Maslin (both scientists from the University College London) conclude that a business-as-usual approach—in the sense that societies will show resilience to adapt to climatic changes—cannot continue because of negative feedback loops that will cause economic collapse. This is their stark conclusion:

> Given that long-run economic growth, powered by fossil fuels, is central to today's global economy, and painless technological fixes are eluding us, it is not clear how climate change and collapse can be avoided without changes of policy or some rupture to today's mode of living. Such difficulties mean that many experts worry that carbon dioxide emissions will accumulate, leading to 3°C warming above pre-industrial levels, even after accounting for the pledges of counties in the Paris Agreement. Higher levels or perhaps even 4°C could occur if not all pledges are met. This level of warming would include risks that are seriously difficult for society to manage. It would quite literally change the world we live in.[58]

One may observe that it is rather odd that a basically conservative business-as-usual approach would carry such severe risks, given the assumptions that the Earth system is robust, that a market-oriented economic system will be able to solve environmental problems, and that scarce resources are better allocated to address pressing problems such as scarce water, sanitation, health, and education.[59]

Pope Francis may serve as a final witness. In his Apostolic Exhortation *Evangelii Gaudium* (2013), the recently elected pope responds to the economic inequality and financial meltdown by saying "No to an economy of exclusion":

> 54. In this context, some people continue to defend trickle-down theories which assume that economic growth, encouraged by a free market, will inevitably succeed in bringing about greater justice and inclusiveness in the world. This opinion, which has never been confirmed by the facts, expresses a crude and naïve trust in the goodness of those wielding economic power and in the sacralized workings of the prevailing economic system. Meanwhile, the excluded are still waiting. To sustain a lifestyle which excludes others, or to sustain enthusiasm for that selfish ideal, a globalization of indifference has developed. Almost without being aware of it, we end up being incapable of feeling compassion at the outcry of the poor, weeping for other people's pain, and feeling a need to help them, as though all this were someone else's responsibility and not our own. The culture of prosperity deadens us; we are thrilled if the market offers us something new to purchase. In the meantime all those lives stunted for lack of opportunity seem a mere spectacle; they fail to move us.[60]

In his encyclical *Laudato Si'* Pope Francis again emphasizes limits to economic growth. The recognition of limits runs like a thread throughout the encyclical. The word "limit" appears in various forms no less than forty-nine times, while in/finite appears seventeen times in the text of the encyclical. The pope critiques the vices of mastery and exploitation, the inability to set limits to our needs (§11), the limitless expansion of consumer goods (§34), the utilitarian mind-set that emphasizes individualism, competition, consumerism, and an unregulated market (§210), the modern myth of unlimited material progress (§78), the lie that there is an infinite supply of the Earth's goods (§106), the vices of self-centeredness (§204[61]) and a limitless mastery over everything (§224), indeed the "cult of unlimited human power" (§122), and the technocratic paradigm that made it "easy to accept the idea of infinite or unlimited growth, which proves so attractive to economists, financiers and experts in technology" (§106). He recognizes three aspects of such limits to economic growth, namely, the assumption that "an infinite quantity of energy and resources are available, that it is possible to renew them quickly, and that the negative effects of the exploitation of the natural order can be easily absorbed" (§106). In response, Pope Francis stresses the need to direct, develop, and limit human power (§78). It is particularly important to note his emphasis on the virtues of humility, frugality, self-restraint, and equity. If there is to be room for healthy economic growth in some parts of the world, this has to be enabled by restrictions elsewhere:

> In any event, if in some cases sustainable development were to involve new forms of growth, then in other cases, given the insatiable and irresponsible growth produced over many decades, we need also to think of containing growth by setting some reasonable limits and even retracing our steps before it is too late. We know how unsustainable is the behaviour of those who constantly consume and destroy, while others are not yet able to live in a way worthy of their human dignity. That is why the time has come to accept decreased growth in some parts of the world, in order to provide resources for other places to experience healthy growth. Benedict XVI has said that "technologically advanced societies must be prepared to encourage more sober lifestyles, while reducing their energy consumption and improving its efficiency" (§193).

Remarkably, this emphasis on discerning limits is enabled by a recognition of the infinite, the infinite distance between Creator and creature (§88), especially God's infinite wisdom, goodness, love, and beauty (§69, 243). It is difficult to recognize such limits if we consider ourselves autonomous, when we exclude God from our lives (§224). By contrast, mystic experiences of the intimate connection between God and all beings allow for a recognition of their goodness (§234). It is the idolatrous attempt to reach beyond human limits, to become like God, that lies at the root of the unwillingness to accept limits to economic growth. This is a theme that will be explored in more depth below.

ON HUMANISM: THREE PILLARS

The second candidate for an ideology that is distinctive of the Anthropocene is given in the name, namely, the humanism that forms an integral part of modernity but is certainly also deeply embedded in African philosophy and in postcolonial and decolonial discourse. Such humanism is built on the three pillars of human exceptionalism (and therefore supremacy?), human dignity, and human responsibility. The distinction between what is human and what is inhumane is indeed an ethical assumption of humanism.[62] These features are widely taken for granted in (Western) ethical theories but have become much disputed in literature on the Anthropocene. Virginie Maris captures the problem in terms of the human reciprocity assumed in moral awareness:

> Humanism and the consonant anthropocentrism are deeply rooted in reciprocity. Western traditional moral theories all depend on the recognition in the other of the same kind of fundamental interests or the same inherent autonomy as those of the moral agent. We are looking for ourselves in the others in order to grant them our moral consideration. This narcissistic feature of humanism, if able to ground morality inside quite homogeneous cultural groups like white enlightened Western men, has failed to create the genuine moral framework needed in the pluralist and globalised societies we now live in. We urgently need to find ways of rethinking morality from a different perspective. The recognition of the otherness of the others and the respect of differences, rather than the desperate seeking for resemblances could be fruitfully stimulated by the reconsideration of human-nature relationships that are grounded neither in assimilation nor rejection.[63]

Put differently, can one emphasize human responsibility without an appreciation for human dignity? And can one emphasize human dignity without an appreciation for the intrinsic worth of each human being and all human beings? On what is such dignity based? This is tricky not least because of attempts to avoid the religious rationale, namely, being created "in the image of God" which presumably does not apply to other species. To base a sense of human dignity on any specific human characteristic (rationality, language, symbolic behavior, labor, artistic ability) exacerbates the problem of using anthropocentric criteria. This is not innocent since human eating patterns are based on such distinctions. Why else would a society condone eating chickens but not (human) children? Or why would vegetarians condone eating chickpeas but not chickens? Typically, it is the intrinsic worth of humans that is used as criterion to guide cultural patterns. If the intrinsic worth of other forms of life is acknowledged, then the *equal* intrinsic worth cannot be affirmed without calling into question such eating patterns. If intrinsic worth is graded (which is the obvious solution), this again raises the question what criterion is used for such gradation. Complexity? Does such a criterion not typically remain anthropocentric too?

Moreover, once a graded intrinsic worth is acknowledged to distinguish between species, the question emerges why such gradation is not applied among human beings too. How, then does one address problems regarding infants, the severely mentally handicapped, the senile, the comatose, the terminally ill? In practice, human worth is of course graded (with salaries, privileges, benefits, and burdens to match that), but legal systems in democratic societies depend on an affirmation of equal human dignity. The inalienable dignity of each and every human being, irrespective of gender, race, rank, creed or ability, is the cornerstone of any bill of human rights. This cannot be graded without bringing society in jeopardy.

Such a notion of equal human dignity rests upon a Cartesian and Kantian universalism, namely, that all human beings share a common rationality that also distinguishes humans from other animals. However, such a (hypothetical) universalism has been steadily undermined through the various "turns" in the humanities and social sciences over the past century or so, including the linguistic, hermeneutical, cultural, sociological, gendered, and spatial "turns." These turns have yielded a deeply held skepticism about such universalist assumptions and widespread recognition of difference more than similarity, an ethical appreciation for diversity, otherness, marginality, liminality, hybridity. The fundamental objection is that the very idea of "Man" often serves as ideological cover for imperialist or colonialist domination by race, gender, culture, sexual orientation, class, and now also species.[64] This can easily degenerate into a consumerist conformity.

Nowadays such a notion of (hypothetical) universalism can also be based on genetic continuity within the human species (despite small differences in the presence of Neanderthal and Denisovan genes). This helps to guard against crude forms of racism which often relies on some form of social Darwinism. But if genetic similarity is the criterion, how should the 98+ percent continuity between humans and chimpanzees be assessed? Or the 60 percent similarity with banana trees? Admittedly, 1 percent may make a huge difference, for example, between a leaking bucket and one holding water. Likewise, there is a huge difference between 350 and 450 parts per million of carbon dioxide in the Earth's atmosphere (a 0.01 percent difference). What would therefore count as sufficient genetic continuity to acknowledge equal intrinsic worth?

The way out of this dilemma may be contract theory (whether based on Hobbes or Rousseau or their followers), namely, to assume that such dignity is based on a long-standing social contract, that is, as an intraspecies rule (based on the will of the majority?). If so, the question becomes under what conditions this contract can be annulled. Would apocalyptic scenarios around societal collapse due to climate change allow for changes to such a social contract? Is what is required not something different, perhaps the kind of "natural contract" that Michel Serres proposes, a peace treaty to put an end to the war against nature?[65]

HUMANS IN THE CENTER? FOUR CONTRASTING POSITIONS

The question is not whether we humans are part of the ecosystems that we inhabit but what our place and role within it may be. Any rigid divide between nature and culture as assumed in classic forms of humanism can no longer hold. An emancipation from the harshness of nature by researching the laws of nature is in the end not possible.[66] In the Anthropocene this has become a question raised across many disciplines.[67] As many observers have wondered, the question is whether the human species operate more like a colossal cancerous growth or like a nervous system able to pick up, contemplate, and respond to pain impulses. Do we add value richness to biotic systems through culture (like organic fertilizers) or do we extract what is valuable (through mining) so that in the end we diminish value? Are we in the center of things that surround us (the "environment")? Or are we living within them like parasites?[68] The question has an even wider scope: where is the history of hominins in general and of our species in particular heading toward? How does human history relate to planetary history? Did humanism not make us the exact midpoint and excellent culmination of all things?[69] What is the significance of this point in the coevolution of humans and other forms of life?

If I am not mistaken, given the impact of human societies on Earth systems, four diverging positions on the place of the human species within evolutionary history have emerged in literature on the Anthropocene. These are (1) the abandonment of humanist and modernist assumptions through various forms of posthumanism; (2) in response to that, a defense of Enlightenment humanism in the form of ecomodernism that unabashedly affirms the liberating impact of science and technology and calls for a continuation of the modern project to address its side effects; (3) the more ambitious proposal for transhumanism with its emphasis on genetic engineering, human enhancements, and artificial intelligence; and (4) those like Clive Hamilton, for whom it is the entire human project that is at stake in the Anthropocene and are therefore not shy in calling for a "new anthropocentrism." These contrasting positions that emerged in critical altercation with each other indicate the heart of the ideological debate on the Anthropocene and cannot readily be captured in a few paragraphs. In the discussion below, I will sketch the first three positions in broad outline with reference to one recent contribution in each case and will then discuss Hamilton's position in more detail.

a) The intuition behind posthumanist positions is that the modern turn to the subject exacerbates an underlying anthropocentrism in the Western tradition that has been religiously legitimized by the Abrahamic faiths that introduced a clear distinction between Creator and creature, desacralized the world, and declared the human species to be the "crown of creation." This follows Lynn White's provocative essay in which he maintained that "in its Western form, Christianity is the most anthropocentric religion that the world has ever seen."[70] Some suggest that human exceptionalism is the root of the problem and that monotheism has served to undergird such assumptions, namely, to explain what is so special about humans and why humans may dominate other forms of life. Here the critique of modernity requires a critique of religion, including theocentric positions.[71] It is the pride of anthropocentrism that is the root cause of the disenchantment and the subsequent alienation from nature. The diagnosis of posthumanist approaches is therefore that anthropocentricism is the ideology that lies at the root of ecological destruction.

This critique is extended also to secular forms of humanism. At the core of modernity is an affirmation of human uniqueness, inviolable human dignity, and human responsibility to rule over all other species. The critiques of domination in the name of differences of gender, race, and class are thus extended to domination in the name of differences of species. The attempt to create a more humane society, even the humane treatment of other animals, does not address the underlying problem. It is to treat other animals as if they were human. In the words of Yuval Harari: "The founding idea of humanist religions such as liberalism, communism and Nazism is that *Homo sapiens* has some unique and sacred essence that is the source of all meaning and authority in the universe. Everything that happens in the cosmos is judged to be good or bad according to its impact on Homo sapiens."[72] Indeed, humanism is itself the dominant world religion and attempting to fulfill the humanist dream is likely to cause its disintegration.[73]

In response, posthumanist positions advocate that humans have to rediscover their proper place in the "community of life" and in the evolution of species. The human species arrived late on the scene in terms of biological and especially cosmic evolution. It pales into insignificance in the vast expanses of space and time. Accordingly, humanism is part of the problem and not of the solution.

There is a wide array of alternatives to humanism that are explored in posthumanist discourse. In environmental ethics such radical alternatives include biocentrism, ecocentrism, deep ecology, and ecofeminism while social ecology allows some hierarchical distinction between first nature and second nature, between human civilization and forms of animal sociality. Typically, the critique of anthropocentrism and alienation from the Earth is countered by an emphasis that we humans belong to nature, that we are "at home on earth." One pertinent example of such a critique of human exceptionalism is provided by Donna Haraway although she resists the posthumanist label. In *Staying with the Trouble: Making Kin in the Chthulucene* she draws on her work in the field of biology to describe the intricate web of life where any one specimen relies on numerous others to stay alive. "Making kin" is what is required on the basis of this recognition. She argues that we human beings are best understood as being compost (humus) that can enrich other forms of life.[74]

To state that humanity forms part of a web of life is easier affirmed than thought through. Is the very longing to be at home not based on an infantile longing for a harmony secured by a divine Creator?[75] As Bruno Latour astutely observes, the Western tradition has insisted that humans be distinguished from the rest of nature. He says: "This is what is meant, most often, by the notions of culture, society or civilization. As a result, every time we attempt to 'bring humans closer to nature,' we are prevented from doing so by the objection that a human is above all, or is also, a cultural being who has to escape from nature, or in any case be distinguished from, nature."[76] In response, Latour recommends "facing Gaia," a position that seems to assume some distance from the face of Gaia.

b) The posthumanist critique of humanism has itself prompted a critique that may be captured under the rubric of "ecomodernism." The argument is that the posthumanist critique is mistaken. It throws the proverbial baby out with the bathwater. It is reductionist in the sense that the value richness introduced by levels of sentience, consciousness, self-consciousness, and symbolic consciousness is not recognized. This does imply a hierarchy, but this can hardly be avoided and is the basis for distinctions between what is mineral, plant, animal, and human. The evolution of life on Earth has a tendency toward increasing levels of diversity and complexity. This should be affirmed. The Enlightenment project of modernism rightly recognized features such as human freedom, autonomy, rationality, and dignity and build modern notions of civilization around that. The fruits of such civilization are there for all to see, namely, advances in the fields of technology, the use of energy, food production, medicine, communication, mobility, democracy, a recognition of human rights and critiques against war, violence against women, exclusion on the basis of race, class, gender, and sexual orientation, and so forth. Such fruits are readily embraced by non-Western cultures.

It may be true that the advent of modernity brought many side effects, including pollution, the exploitation of labor, a loss of biodiversity, and now climate change and ocean acidification. However, each of these challenges has been recognized and is being addressed. This is best done on the basis of the core assumptions of modernity, making use of more advanced forms of technology that would produce more but have less impact. The humanist project should be extended toward its next phase, namely, ecomodernism. It may also be true that modernity was based on slave labor, colonial exploitation, and imperialist rule, but as modernity developed these oppressive aspects were recognized on the basis of the core assumptions of modernity, critiqued and subsequently addressed. The Enlightenment project remains unfinished in the sense that poverty, hunger, a lack of education, disease, gender-based violence, and more localized forms of war are still present. The track record of modernity to address such challenges is excellent and offers confidence that future challenges can be addressed, for example, through programs focused on the United Nations sustainable development goals. The achievement of such goals is being undermined by posthumanist critiques that underplay the value of the Enlightenment. It is also being undermined by religious forms of fundamentalism and a legitimation of narrow group identities. Not surprisingly, ecomodernism is unabashedly secular and often atheist in orientation. There is hardly room for any form of Christian ecomodernism.

In *Enlightenment Now* Steven Pinker promotes such an ecomodernist retrieval of humanism. He puts his trust in human progress through rationality, open and critical inquiry, and technological advances. He gathers together an array of data to demonstrate such progress in numerous fields, including food, health, wealth, equality, war, democracy, and rights. He acknowledges environmental concerns, including climate change and the need for decarbonization, but insists that these are problems to be solved, not insurmountable obstacles. It is possible to decouple human flourishing from the exploitation of natural resources. He then adds this:

> It's time to retire the morality play in which modern humans are a vile race of despoilers and plunderers who will hasten the apocalypse unless they undo the Industrial Revolution, renounce technology and return to an ascetic harmony with nature. . . . The tide of modernity does not sweep humanity headlong toward ever more unsustainable use of resources.[77]

Pinker's emphasis on Enlightenment humanism is unapologetic and at times arrogant in its Western predispositions. His position is also unabashedly atheist although he welcomes forms of spirituality that aligns itself with humanist values. The way in which he constructs the story of human progress is remarkable and his conclusion is worth quoting at some length:

> The story of human progress is truly heroic. It is glorious. It is uplifting. It is even, I daresay, spiritual. It goes something like this:
>
> We are born into a pitiless universe, facing steep odds against life-enabling order and in constant jeopardy of falling apart. We were shaped by a force that is ruthlessly competitive. We are made from crooked timber, vulnerable to illusions, self-centeredness, and at times astounding stupidity.
>
> Yet human nature has also been blessed with resources that open a space for a kind of redemption. We are endowed with the power to combine ideas

> recursively, to have thoughts about our thoughts. We have an instinct for language, allowing us to share the fruits of our experience and ingenuity. We are deepened with the capacity for sympathy—for pity, imagination, compassion, commiseration . . .
>
> These endowments have found ways to magnify their own power. . . . We will never have a perfect world, and it would be dangerous to seek one. But there is no limit to the betterments we can attain if we continue to apply knowledge to enhance human flourishing. The heroic story is not just another myth. . . . And the story belongs not to any tribe but to all of humanity—to any sentient creature with the power of reason and the urge to persist in its being. For it requires only the convictions that life is better than death, health is better than sickness, abundance is better than want, freedom is better than coercion, happiness is better than suffering, and knowledge is better than superstition and ignorance.[78]

There can be little doubt about such human "progress," at least in terms of scientific knowledge, technological innovation, and medicine. This has been the main stimulus behind economic "growth," albeit on the basis of the exploitation of "natural resources." Often notions of "development" are aligned to such ideas of growth with the assumption that progress may be shared with all of humanity, although such development takes place despite increasing economic inequalities and at the cost of other species. Pinker realizes that progress, growth, and development cannot be equated with human happiness or flourishing, but his tacit assumption is that such flourishing is no longer possible without such progress. He also realizes that human flourishing cannot be sustained without ecological integrity, but his notion of humanism suggests that the latter remain in the service of the former. In this regard, he stays true to the Enlightenment ideals that he espouses. How human flourishing relates to the virtues of wisdom, justice, courage, temperance, faith, hope, and love will require further exploration.

c) In the transhumanist critique of humanism some announce an end to humanism, to be replaced by techno-humanism—enhancing human abilities through artificial intelligence, genetic engineering, and nanotechnology so that being human becomes another designer product. With genetic engineering we continue to transform what being human may become at an accelerated pace, not knowing where this will lead to.[79] David Wallace-Wells captures the escapist appeal of Silicon Valley: "the belief that the grandest task before technologists is not to engineer prosperity and well-being for humanity but to build a kind of portal through which we might pass into another, possibly eternal kind of existence, a technological rapture in which conceivably many—the billions lacking access to broadband, to begin with—would be left behind."[80]

Alternatively, humanism will be replaced by the rule of big data—an envisaged Internet-of-all-Things where the flow of information will spread across and from the Earth to pervade the whole galaxy. Either way, in Yuval Harari's version of the history of the future, *Homo sapiens* will be superseded and eventually be replaced by *Homo Deus* in the same way that *Homo sapiens* outperformed, marginalized, and eventually replaced *Homo Neanderthalensis*, probably through better communication and thus cooperation strategies. This may well result, Harari observes, "in the creation of a new superhuman caste that abandon its liberal roots and treat normal humans no better than nineteenth-century Europeans treated Africans."[81] Alternatively, humans are merely the tools for the creation of a cosmic data-processing system. Such a system would be like God: "It will be everywhere and will control everything, and humans are destined to merge into it."[82] Harari adds:

> In the days of Locke, Hume and Voltaire humanists argued that "God is a product of the human imagination." Dataism now gives humanists a taste of their own medicine, and tells them: "yes, God is the product of the human imagination, but human imagination in turn is just the product of biochemical algorithms." In the eighteenth century humanisms sidelined God by shifting from a deo-centric to a homo-centric world view. In the twenty-first century, Dataism may sideline humans by shifting from a homo-centric to a data-centric view.[83]

The word "just" is significant here as Harari's argument is unashamedly reductionist: organisms are nothing but complex algorithms. Admittedly, he does question this assumption himself,[84] so that his argument may partly serve a rhetorical purpose. This is hard to tell. Or is Harari following the Platonic route that ideas have an ontological priority over matter? Are words (divine decrees? algorithms?) ultimately creative, so that life is made of language, so that it is the word that becomes flesh in any form of life?[85]

d) In his book *Defiant Earth*, Clive Hamilton offers a critique of posthumanism, ecomodernism, and transhumanism alike. He proposes, instead, a new anthropocentrism. It may be worthwhile to follow Hamilton's argument in this regard at some length:[86]

Hamilton seeks to fathom the emergence of this species with an ability to alter the functioning of Earth systems. He calls for a "new anthropocentrism" to acknowledge human uniqueness and power, to "instantiate humankind once and for all as the being at the center of the Earth" (41). This is not to be misunderstood as a normative claim that humans should be the masters of the planet but as a descriptive fact, namely, that humans have become the dominant factor in Earth systems, that humans are positioned at the center of the world (43). It is too late to abandon an anthropocentric standpoint as if a biocentric, ecocentric, or theocentric alternative is possible. For Hamilton, "Never has the 'is' been more apparent, never has the 'ought' been less defensible" (43). It is too late to pull back in the hope that Earth systems can revert to a predisrupted state. The disruptions are now irreversible and will persist for millennia (51). For Hamilton, this "descriptive fact" is that human technological power is indeed able to alter the future of the Earth so that we must fear its misuse and must learn to deploy it more judiciously and respectfully (39). Humans may have acquired such power foolishly, but "we now have responsibility for the Earth as a whole and pretending otherwise is itself irresponsible" (43). The question is no longer whether we accept that responsibility, but how we exercise that (44). Our obligation is to accept, to own, and even to embrace an amplified responsibility (53).

This "new" anthropocentrism is distinguished from the "older" versions (whether biblical, Cartesian, Kantian, or Marxist) that humans are entitled to use nature, either through domination and exploitation or through gratitude and stewardship (53–54). Hamilton rejects ideas of domination and of stewardship alike. The core difference between the old (Kantian) anthropocentrism and the new anthropocentrism is that the former regarded a "humanized Earth" as inevitable and desirable, whereas the latter regards it as destructive and ambivalent at best. He explains:

> Whereas the old form declares that humankind has a special place on Earth that gives it moral dominion over nature, the new form points out that humankind has actual power to change the course of the Earth, but draws the opposite ethical conclusion. The old form claims that the Earth provides endless bounty for the sake of humans; the new form believes that the Earth will not provide bounty unless humans take care of it. Instead of concluding that humans are free to do whatever we want on a passive Earth, the new anthropocentrism insists that on an active and fractious Earth humans are not free to do whatever we want but must restrain ourselves and restrict what we do. The old form was humanist because human destiny was thought to lie in human hands; the new is an anti-humanist anthropocentrism because the destiny of humans now lies as much in the hands of "Gaia" as in our own. . . . The new anthropocentrism puts itself outside the usual debate over values, including the notion of intrinsic value. It is not a question of what does and does not have value; it is an approach to the whole. (54, 55)

The new anthropocentrism is therefore based on a twofold recognition, namely, of humans as world-makers but also that the material conditions within which such world making is situated, from an Earth systems perspective, now constitute the Anthropocene. If human existence is situated, it is now situated in the Anthropocene and that applies to all cultural contexts.

Hamilton suggests that such an emphasis on world making avoids the biologism of species talk and the mystification that surrounds grand narratives of cosmic evolution that comes to fruition in human creativity (65). We humans are "capable of enormous creative renewal but equally capable of catastrophic hubris and overreach" (66). The world that we constructed is both sustaining us and destroying us.[87]

Hamilton distinguishes such a new anthropocentrism not only from the "old" (Kantian) anthropocentrism but also (on the one hand) from the technological optimism of ecomodernists and (on the other) the pessimism of posthumanists.

He argues that ecomodernists expect the coming of a great Anthropocene, one allowing for universal human dignity and a biodiverse and thriving planet, with the liberal principles of democracy, tolerance, and pluralism as the key driving principles to achieve that.[88] Accordingly, progress is driven by human creativity. It is progress that will ensure that the good Anthropocene prevails. Indeed it operates as a functional replacement for God's benevolence (69). This spawned an anthropo-supremacism, and ensures our sense of domination over the Earth and indeed over other humans. Hamilton is not shy of praising human creativity in the form of technology, but also recognizes that technology embedded in social structures presents the greatest danger (e.g., in the form of geoengineering) for human flourishing (73). Even as an atheist he recognizes creativity as a divine gift that could be used to separate ourselves from God and usurp God's power, but agrees that the gift may also be used in humility and gratitude in a morally responsible way (74–75).

Hamilton's critique of posthumanism is subtle and significant. He acknowledges that posthumanism has been the dominant trend in the humanities and social sciences, at least since the 1960s and has managed to deconstruct social hierarchies of power and control:

> Feminism set out to cut men down to size; queer theory destabilized heteronormativity; and postcolonial and subaltern studies wrote "history from below" to delegitimize the stories from above. They worked by holding humanism to account, by turning its principles of equality, human dignity and justice against the societies that proclaimed them. (87)

Hamilton observes that this social critique was then extended to address human domination of nature. Anthropocentrism became the focus of this critique—beyond Eurocentrism, androcentrism, and heteronormativity—deconstructing modernity's narrative of human domination of the Earth by questioning the nature/culture binary. Anthropocentrism is regarded as an attitude, one that "present the human species as the centre of the world, enjoying hegemony over other beings and functioning as masters of a nature which exists to serve its needs" (90).[89] Accordingly, the Enlightenment view of nature is inextricably tied to European colonial domination. Hamilton describes this posthumanist extension as an unwarranted epistemological leap (87). His argument is that the arrival of the Anthropocene precisely indicates the power of humans to influence Earth systems. Contrary to the posthumanist stance, the task of the humanities now is not to undermine such power but to recognize and accept the unique and extraordinary power that humans have in order to exercise that responsibly (87). The posthumanist tendency is to deflate the significance of human power by insisting that human separation from nature is delusional in order to symbolize the failure of the modernist project of domination (89–90). It undermines human uniqueness in order to emphasize interspecies relationships (93). According to new materialism, humans are shaped by matter more than the inverse so that destructive human agency can become deflated (94–95). By contrast, Hamilton insists that the Anthropocene elevates rather than deflates human power. At the very historical juncture where the power of humans has become more evident than before, it seems that posthumanists wish to dethrone human power and deprive humans of their uniqueness. No other force (living or dead) has the capacity to influence the Earth system *and* to decide to do otherwise (99). He captures this vividly:

> While the urge to counter human arrogance and misuse of human powers is necessary, the effect of this kind of reduction is to decentralize causality at a time when power has in reality never been more centralized. In this way, anti-anthropocentrism has the perverse effect of denying our responsibility for the damage that we have caused. (98)

On this basis, Hamilton sets out to defend human uniqueness so that the human story is recognized to be the most remarkable story ever told (92). Humans stand out from nature as a whole (99). His argument seems to be based on the destructive impact of human power. It would not do to regard such power merely as a form of social construction. He acknowledges the posthumanist defense of distinct knowledges (ways of seeing the world) and ways of being in the world, but argues that a particular ontology, namely, Western naturalism, has become so utterly dominant that it continues to marginalize other positions (animism, totemism, and analogism) so that these positions need to redefine themselves in dialogue with modernism, especially in the sprawling cities of Asia, Latin America, and Africa. Their worlds are being sucked into the Anthropocene. Ways of addressing the Anthropocene cannot rely on such premodern worldviews because they do not and cannot account for technological abilities or for their destructive impact. The only way forward is to begin with the impact of modernity in order to move beyond modernity (106). He says: "There is no going back to pre-modern ontologies for an understanding; we must look ahead to the evolution of modernity itself, driven by its own endogenous forces and contradictions within a larger order" (110). Modernity yielded the power both to destroy and to protect the Earth so that the main question now is how that power will be used.

Hamilton's argument for a "new anthropocentrism" is not only critically juxtaposed to other positions but also constructive. It recognizes the significance of the emergence of humanity in the evolution of species. What kind of species is this that is able to wield such power? Instead of cutting humans down to size by insisting on interspecies dependence in a flatland of entanglement, he suggests that the Anthropocene indicates that humans are not just another species (95). In a move reminiscent of Heidegger, Hamilton argues that humans "stand out" from nature as a whole (99). Humans are indeed embedded in nature, but this both constrains and allows for human agency and subjectivity. The Anthropocene thus requires an anthropology that does not deprive unique (i.e., intentional) human agency but recognizes human distinctiveness within networks (101). The greatness of the modern project lies in combining freedom from oppression with power over nature through using science and technology and the institutions that mobilized them. It is not only our contemplative ability that characterizes the human species but our ability to transform the Earth (119). The task that confronts humanity now is to use such power to pacify and protect the Earth in order not to destroy it (110–11).

Even if human existence pales into insignificance in terms of cosmic space and time, the real prospect of human extinction cannot be trivialized: "For it is we who give the Earth meaning and mark it out as the unique planet in the cosmos" (114). Hamilton resists the cold acceptance of the destruction of humankind's "immense cultural and intellectual accomplishments" (118). He regards attempts to instill a sense of humility by cutting humans and indeed the human project down to size as "the gravest moral error" because this project is the Earth's greatest project (114). The Earth is the planet of humans. He resists the posthumanist characterization of humans as "only one of very many species and not obviously worth preserving" (113, with reference to a formulation by John Gray), "mere compost to be reabsorbed," a "mere assemblage of molecules that came together for a short time to resist entropy" (117). He regards such positions as irresponsible: "Fatalistic acceptance of human extinction is common among post-humanists whose understandable disgust at the brutality of ecological destruction delivers them into a quietism rooted in disavowal of the human project as a whole" (115).

Hamilton concludes that extinguishing humans from the Earth would be a tragedy of cosmic significance (116). Inversely, if human existence is meaningless, then humankind's destruction of the conditions for life on Earth to flourish is also meaningless (117). Hamilton's "new anthropocentrism" thus returns to a Kantian emphasis on the unique moral worth of this species. What is being destroyed is the human lineage. What kind of species is this that is able, even destined, to destroy itself? If the tendency to transform its own environment is inherent to the species, then the species as such (and not only capitalist culprits) contains within itself a destructive edge. This suggests a tragic vision, if not an underlying Manichaeism.[90]

I have serious reservations as to whether the word anthropocentr*ism* is best used for the "descriptive fact" of human mastery. For Hamilton the problem is that we have not accepted responsibility for the power that we do possess which leads to a "monstrous anthropocentrism"—elevating the human to a position of power without responsibility (43). That point is well taken but anthropocentrism then seems to be a misnomer that can only create unwanted terminological confusion, not least by suggesting that "we are not anthropocentric enough" (53). It also does not help to distinguish between an arrogant and a humble anthropocentrism (44). Hamilton finds a split of human subjectivity from nature a necessary precondition for freedom and maturity. Although this opens the possibility of the destruction of nature, this is not necessary and also opens up the possibility of resistance to such destruction (143).

REFRAMING THE HUMAN

It seems at least clear that with the advent of the Anthropocene the classic question of human self-understanding received a new twist: what does it mean to be human in the age of humans? Almost every scholar engaged in reflection on the Anthropocene from the perspective of the humanities has recognized the need to rethink the place and role of humans within the biosphere, and to assess the cosmic significance of the emergence and history of this one species. Just at a time when the concept of being "human" seems to become outdated and being replaced with posthumanism, the "Anthropos" has come back with a vengeance.[91] The question why the world is like it is (and what has gone wrong with it) poses the need for a reassessment of the place of humans within evolutionary history.

An influential essay by Dipesh Chakrabarty entitled "Postcolonial Studies and the Challenge of Climate Change" may provide a way forward. He identifies three distinct images of the human (besides anti-humanist and post-humanist views): "The universalist-Enlightenment view of the human as potentially the same everywhere, the subject with capacity to bear and exercise rights; the postcolonial-postmodern view of the human as the same but endowed everywhere with . . . differences of class, sexuality, gender, history, and so on . . . [and] the figure of the human in the age of the Anthropocene, the era when humans act as a geological force on the planet, changing its climate for millennia to come."[92]

Chakrabarty observes that the one need not be rendered invalid or be displaced by the other. One may need to speak of humanities in the plural, but to speak of a common humanity has enabled resistance against racism and sexism. He adds that the human subject in the age of the Anthropocene is distinct from the Kantian emphasis on universality or the postmodern/postcolonial emphasis on difference: "If critical commentary on globalization focuses on issues of anthropological difference, the scientific literature on global warming thinks of humans as constitutively one—a species, a collectivity whose commitment to fossil-fuel based, energy-consuming civilization is now a threat to that civilization itself."[93] Indeed, climate change "calls us to visions of the human that neither rights-talk nor the critique of the subject ever contemplated."[94] It does not make rights-talk or the postcolonial critique of the subject redundant, but it does elicit reflection on the unprecedented planetary scale of human impact on Earth systems and not only local or regional environments.[95] Nevertheless, since there is no corresponding "humanity" that can serve as a political agent to address climate change, and given the uneven impact of climate change, a politics of difference will be required to address intrahuman issues of justice.[96] For Chakrabarty, because the forces of globalization and global warming intersect, there is a need to view human beings simultaneously through contradictory registers: as both a geological force and as political agents in an unequal, unjust world.[97] There is a need to hold onto a universal history of life while retaining a postcolonial suspicion of the universal.[98]

Chakrabarty's willingness to hold together these three contrasting, even contradictory images of being human is at least honest in the sense that it recognizes the unresolved nature of the debate. The tensions are undeniable but not easily overcome. It may suffice to offer two further comments:

First, there is some unnecessary confusion in the literature on the term anthropocentrism. Some suggest that a weak form of anthropocentrism is epistemologically (or hermeneutically) necessary and indeed inevitable: human valuing will always be from a human point of view.[99] This is the core Cartesian and Kantian insight, namely, that the understanding subject is the clue to understanding any object. The Copernican revolution introduced by Kant oddly led to the belief that the object revolves around the subject![100] The legitimacy of the hermeneutical insight that any form of language is necessarily *human* language cannot be denied. The epistemological turn to the subject in modernity suggests that human self-understanding is the center from which humans necessarily begin. Or, as Tony Kelly suggests, "the human is not a centre in which we end, but a centre from which we begin;

not self-centred fixation, but the self-transcendence into the other, the more, the whole."[101]

However, a conceptual distinction may be appropriate here, namely, between anthropogenic, anthropomorphic, and anthropocentric categories to assess the place of humans.[102] If the use of anthropogenic categories by humans is unavoidable, and some anthropomorphisms can therefore be condoned, anthropocentrism is quite another matter. As Holmes Rolston suggests, nonsentient organisms "are valued, when humans encounter them, for what they are in themselves, and not just for the sake of human appreciation."[103] With Baird Callicot one may distinguish between tautological anthropocentrism (human valuing is inevitably a human act), moral anthropocentrism (recognizing the intrinsic worth of all human beings and only human beings), and metaphysical anthropocentrism (the conviction that humans occupy a privileged place in the order of being).[104] The first is trivially true; the second and third do not necessarily imply each other. Recognizing the intrinsic worth of nonhuman forms of life does not imply equal intrinsic worth either. Value richness can still be graded.

Second, the age of the Anthropocene entails two contrasting movements, namely, a recentering of the role of humans in planetary history, following the apparent decentering associated with the Copernican and Darwinian revolutions. In the light of cosmic and biological evolution, humans can hardly be regarded as the center of the universe, or as the hinge upon which history turns. We are latecomers in the evolution of species on Earth, while the Earth itself pales into insignificance given the vast expanses of space and time.[105] To place the Earth in the center reflects our human narcissism, the form of humanism that renders us "the exact midpoint or excellent culmination of all things."[106] Yet, such a decentering of the human was itself only possible through the Cartesian and Kantian turn to the subject and the revolutionary emphasis on freedom, equality, and fraternity. Indeed, one may argue that modernity may be associated with an anthropocentric turn away from a theocentric or ecclesiocentric worldview that dominated the period before that. And others would say that this anthropocentric turn remains incomplete given tendencies to replace God or the church with Science, the Market, or the Party.[107] Or is modernity merely the logical outcome of the anthropocentrism embedded in the Jewish–Christian tradition (as Lynn White among others would argue)? Either way, one may argue that the anthropocentric turn in modernity was undermined by its own outcomes in terms of cosmological, geological, biological, and paleoanthropological insights. The Cartesian motto of mastery over nature did not distinguish us from other animals (*cogito ergo sum*) but reinforced an underlying possessive animality.[108] Nevertheless, after centuries of increasing cosmic insignificance, in the Anthropocene humans seem to be back at the center.[109]

The traditional predisposition that human communities are finite but that the Earth seems infinite is reversed. Now the planet seems finite (with implied planetary boundaries and limits to growth), while the capacity for human ingenuity seems infinite. As Michael Northcott observes, "But if the Copernican turn decentered humanity from the cosmos, and reduced the perception of human influence over the Earth and the skies, the Anthropocene is a second Copernican turn because it puts humanity back into planetary history as its most influential shaper."[110] Sallie McFague, likewise, suggests that both a decentering and a recentering of the place of human beings in planetary systems are required. She explains:

> We need to move beyond democracy to biocracy, seeing ourselves as one species among millions of other species on a planet that is our common home. . . . We are profoundly interrelated and interdependent with everything living and nonliving in the universe, and especially on our planet. . . . At the same time we have become, like it or not, the guardians and caretakers of our tiny planet. . . . We are the responsible ones, responsible for all the rest upon which we are so profoundly dependent.[111]

In the Anthropocene such a recentering is partly born from a recognition of destructive human power, that humans have become a geological force of nature. We have become the dominant species on Earth, appropriating the whole world to ourselves, dumping our garbage all around us (in *our* environment), also outside the atmosphere, around the globe.[112] The Dutch ethicist Gerrit Manenschijn argues that the idea that humans exist in partnership with the rest of the Earth community may be too romantic, since it tends to minimize the position of power which humans have come to occupy.[113] Instead, it will be more honest and helpful to acknowledge the extent of human power and to focus then on the moral questions about the responsible exercising of such power. In the words of Philip Hefner already a quarter of a century ago, "We now live in a condition that may be termed technological civilization. This condition is characterized by the fact that human decision has conditioned virtually all of the planetary physico-biogenetic systems, so that human decision is the critical factor in the continued functioning of the planet's systems."[114]

Such a recognition of (destructive) human power prompts the question where such power is derived from. What makes humans so uniquely powerful? What is the significance of a recognition of such human uniqueness, contested as it may be? This is the question that energizes debates on the anthropic principle, a reconstruction of hominid evolution, philosophical theories of mind, and legal debates on human dignity alike. Many answers are possible to such questions but my impression is that there is nevertheless an emerging consensus in the fields of paleoanthropology and animal ethology, namely, that symbolic communication is a distinctive feature of the human species, unparalleled elsewhere.[115] Where this ability is derived from is still open to contestation: Did language come from nature or from culture?[116] Did brain development lead to language or did language lead to brain development. Either way, this symbolic capacity, the ability to represent the world in signs, can be put to different ends by pushing beyond the limits of that world, including the capacity to destroy the world around us. As Elizabeth Kolbert observes, "A tiny set of genetic variations divides us from the Neanderthals, but that has made all the difference."[117] We cannot escape from ourselves simply by decrying anthropocentricity. Discourse on the emergence of complexity has reversed the emphasis on human insignificance in terms of space and time. If complexity is the criterion and if human brains and therefore human societies are the most complex feature of the universe that we know of, then we are arguably back in a pre-Copernican world where the whole history of the universe seems to revolve around (better: culminate in) the emergence of such a species.

In short, then, if the contrasting movements toward both a decentering and a recentering of the place of humanity in cosmic and evolutionary history are recognized, then it is hard to know what to make of humanism as the dominant ideology in the Anthropocene. Is a retrieval of humanism part of the underlying problem or also crucial for the way forward? That clearly depends on what form of humanism is profiled, given the marked differences between traditional precolonial African forms of humanism, a pre-Copernican Renaissance humanism, Enlightenment humanism, ecomodernism, or postcolonial/decolonial forms of humanism. It should at least be clear that Renaissance humanism was unable to define the human given its Greco-Latino rootedness. The divergence of early humans from Africa, spreading over the whole world over the last 100,000 years, has become partially reversed through technologically mediated contact in the global village so that there is now more than ever before a need to redefine what it means to be human.[118] Even so, the electronic catholicity of the internet excludes some as it includes others.[119]

My sense is that these two contrasting movements are derived on two equally important insights that are already captured by Immanuel Kant in the famous conclusion to his *Critique of Practical Reason* (1788):

> Two things fill the mind with ever new and increasing admiration and awe, the more often and steadily we reflect upon them: *the starry heavens above me and the moral law within me*. . . . The first perspective of a countless multitude of worlds as it were annihilates my importance as an *animal creature*, which must give the matter out of which it has grown back to the planet (a mere speck in

> the cosmos) after it has been (one knows not how) furnished with life-force for a short time. The second, on the contrary, infinitely elevates my worth, as an *intelligence*, through my personality, in which the moral law reveals to me a life independent of animality and even of the entire world of the senses, at least so far as may be judged from the purposive determination of my existence through this law, which is not limited to the conditions and boundaries of this life but reaches into the infinite.[120]

One is left with the same two contrasting insights: On the one hand, there is the recognition that human beings form part of nature and that human life is rooted in what is mineral, vegetative, and animal. On the other hand, there is the recognition that being human includes dimensions that transcend being animal (expressed imprecisely by traditional notions of "soul" or "spirit") so that reductionism therefore needs to be resisted. For Michel Serres the "incandescent," the emergence of the humane, is an unfinished project; its glowing should not be extinguished prematurely.[121] Yet, if being human can be regarded as "higher" than being mineral (at least in terms of complexity), then being mineral is more basic than the other three. These "higher" dimensions include the discernment of what is and what ought to be. Likewise, human beings must recognize their finitude and face its consequences in terms of fragility, adversity, and mortality. As Serres observes, "To deny this finitude, starting in childhood, is to nurture unhappy people and foster their resentment of inevitable adversity."[122] At the same time, humans must nurture their sense of the infinite, the thirst for learning, for the beyond, the not yet, for a new world that is possible. The same applies to the recognition of contingency that disturbs our ability to assign suffering either to moral evil or to natural evil (or for that matter divine providence) and the quest to discern pattern, rationality, purpose. Can what is contingent and what is rational be reconciled? How does one make sense of the contingent if it is indeed contingent?[123]

The marker of the Anthropocene indicates that these pairs, let us say being human and being animal, are no longer in touch with each other. Any form of humanism that subsumes the one under the other will become destructive.

ANTHROPOCENTRISM AS CORE DIAGNOSTIC?

The critique of ideology does raise a question of diagnosis: what has gone wrong with humanity that could have such a geological impact? Is the problem pride (anthropocentrism), greed (consumerism), sloth (the failure to accept responsibility), pollution, corruption, folly, or alienation from nature?[124] Or is the problem perhaps mastery in the sense of domination in the name of difference through strategies of exclusion, appropriation, and exploitation? In that case one may argue that the underlying problem is still one of anthropocentrism in the sense that the significance of the differences that are undeniably there—between minerals, plants, animals, and humans—have been misunderstood. This leaves the question whether all humans are complicit in such anthropogenic destruction and to what extent considerations in terms of geography, race, gender, class, language, and culture play a role. In climate change discourse it is widely accepted that there is a common but differentiated responsibility that includes all humans. Even where the "common" culpability and responsibility is disputed, it is at least clear that there is a common predicament.[125] Is anthropocentrism then the appropriate term to diagnose the underlying problem? How does a recognition of the Anthropocene alter (sustain, radicalize, or trivialize) this diagnosis, namely, that it is anthropocentrism (and therefore humanism?) that forms the root cause of the disease?

These are not the kind of questions that can be addressed cursorily. I do need to note that these are also deeply theological questions that I will return to in the postscript. It is nevertheless remarkable to see how secular critics grapple with the problem. Let me call four secular witnesses to the stand on the need for an adequate diagnosis:

In his *Defiant Earth*[126] Clive Hamilton comments that "the primal sin is not to defy God's command but to misuse the gift of our special agency by violating the Earth, destroying the home provided to us" (126). The Earth gave birth to a stubborn *anthropos* who was convinced that it could break all bonds and lord it over the matrix that generated it (140). One may say (with Schelling) that it was the split of the thinking subject from nature that provided the theoretical justification for the destruction of nature (142). However, Hamilton argues that if the split cannot be avoided, the form that it took is at issue: "In the world come of age, it is not our split from nature that must be overcome, but our violence against it, arising from the conviction that the split was a total severance" (143). The split allowed for the possibility of destruction but also for resistance against such destruction. Human agents had a choice and still do. Our coming of age—where the gods had withdrawn and humans "inherited the Earth" (not the meek!)—was the manifestation of an agency latent in nature: "Freedom with Earth-changing power put us in the position where we had to decide how humans could flourish without destroying the Earth" (144).

In my view this diagnosis is remarkable. From a theological perspective it questions sin as alienation and focuses on sin as domination (patriarchal, colonial, and racialized violence) and as folly, while underplaying sin as pride or greed. It does recognize sin as sloth, that is, as the failure to accept the responsibility. It also offers a pervasive notion of sin as encapsulating self-destructive power, capturing a situation from which we humans cannot escape, to which we all contributed (but not equally so) and under which we all suffer (again not equally so). It seeks to fend off the dragon of nihilism (that "stalks every society once it attains a moderate level of affluence and peace" (145)) through the prudent duty to care for the Earth as the species that shapes the face of the Earth.

In an essay entitled "Love your Monsters" published in the ecomodernist *Breakthrough Journal* Bruno Latour goes in a different direction. For him the "monsters" that we have created (the reference is to Dr. Frankenstein) is a cryptic expression for technology. The background is the dialectic of the Enlightenment critiqued by the Frankfurt School where human emancipation from nature has become reversed by becoming enslaved by such technological artifacts. For Latour the answer cannot be to discard such technology but to accept responsibility for that. If anything, the sin is not that of mastery but of sloth. His complex rhetoric needs to be quoted in full:

> The Christian God, at least, is not a master who is freed from dependents, but who, on the contrary, gets folded into, involved with, implicated with, and incarnated into His Creation. God is so attached and dependent upon His Creation that he is continually forced (convinced? willing?) to save it. Once again, the sin is not to wish to have dominion over Nature, but to believe that this dominion means emancipation and not attachment.
>
> If God has not abandoned His Creation and has sent His Son to redeem it, why do you, a human, a creature, believe that you can invent, innovate, and proliferate—and then flee away in horror from what you have committed? Oh, you the hypocrite who confesses of one sin to hide a much graver, mortal one! Has God fled in horror after what humans made of His Creation? Then have at least the same forbearance that He has.
>
> The dream of emancipation has not turned into a nightmare. It was simply too limited: it excluded nonhumans. It did not care about unexpected consequences; it was unable to follow through with its responsibilities; it entertained a wholly unrealistic notion of what science and technology had to offer; it relied on a rather impious definition of God, and a totally absurd notion of what creation, innovation, and mastery could provide.[127]
>
> So here it is sin as sloth, an unwillingness to accept responsibility, the failure to reach the goal (from the Greek *hamartia*) that dominates.

In a public lecture entitled "Decolonizing Knowledge and the Question of the Archive" the Cameroonian public intellectual Achille Mbembe captures the postcolonial agenda of "reframing the human" with explicit reference to the Anthropocene:

> We therefore have to shift away from the dreams of *mastery*. In other words, a new understanding of ontology, epistemology, ethics and politics has to be achieved. It can only be achieved by overcoming *anthropocentrism* and humanism, *the split between nature and culture*. The human no longer constitutes a special category that is other than that of the objects. Objects are not a pole opposed to humans. At the heart of the efforts at reframing the human is the growing realization of our precariousness as a species in the face of ecological threats and the outright possibility of human extinction opened up by climate change.[128]

Mbembe's diagnosis focuses on mastery, but he seems to think that this is the result of a prior alienation from nature, exacerbated through colonialism. It is such alienation that has allowed pride to creep in so that there is now a need to overcome anthropocentrism and humanism—for the sake of human survival.

And here is Vandana Shiva's verdict on the links between alienation and domination:

> How do we turn from the ruins of the culture of death and destruction, to the culture that sustains and celebrates life? We can do it by breaking free of the mental prison of separation and exclusion and see the world in its interconnectedness and non-separability, allowing new alternatives to emerge. . . . We need once more to feel at home on this earth and with each other. We need a new paradigm that allows us to move from the pervasive culture of violence, to a culture of non-violence, creativity, and peace.[129]

I find these secular versions of sin-talk quite remarkable. The full range of traditional (especially Jewish, Christian, Muslim) vocabulary is employed to assess what has gone wrong with the world in the Anthropocene. Sin is alienation, corruption of the good, domination, failure, folly, greed, pollution, pride, rebellion, sloth, name it. Which of these is the primordial sin? That is hard to say. There have been numerous secular and religious attempts to explain one manifestation of sin in terms of another. It is as if one form of sin becomes evident at the surface so that it may be explained by some of the others as the root cause and vice versa. It is perhaps best to see these as deeply intertwined. Sin describes the mess in which we human beings are all caught up, to which we all contributed either as moral agents or moral patients but not equally so, under which we all suffer but again not equally so, and from which we apparently cannot escape because desperate attempts to do so only aggravate the underlying problem.

To conclude, it may suffice to note that the two ideologies discussed here, that is, the assumption that unlimited economic growth in terms of biophysical throughput is possible and the "old" anthropocentrism underlying that, operate like roadblocks that prevent the kind of social imaginary that is required to transform the energy basis of the global economy from fossil fuels to sustainable alternatives rapidly enough.

NOTES

1. If apartheid could be regarded as an ideology, it may best be linked with nationalism. Afrikaner nationalism became the overriding expression of that ideology in response first to Dutch colonialism (the rule of the Dutch East Indian Company) and then British imperialism, irrespective of the interests of Indigenous peoples of Southern Africa. Such nationalism received legitimation in the name of Christianity so that the reigning ideology was often labeled "Christian-nationalism"—prompting endless debates on the status of the hyphen! See, for example, the discussion in Bob Goudzwaard, *Idols of Our Time* (Downers Grove: InterVarsity Books, 1989), 39–48.
2. See Bob Goudzwaard and Craig B. Bartholomew, *Beyond the Modern Age: An Archaeology of Contemporary Culture* (Downers Grove: IVP Academic, 2017), 183–84, with reference to John B. Thompson (*Ideology and Modern Culture* [Stanford: Stanford University Press, 1990]).
3. See Galtung, *Peace by Peaceful Means*, 76. For Galtung, actor conflicts are conscious while structural conflicts are typically not.
4. Ibid., 60.
5. Ibid.
6. Ibid., 65.
7. See Žižek, *Living in the End Times*, 365.
8. See Ibid., 429.
9. On the "unemployment paradox," see Goudzwaard and Bartholomew, *Beyond the Modern Age*, 9.
10. McNeill and Engelke, *The Great Acceleration*, 205.
11. Quoted in Wesley Granberg-Michaelson, *Redeeming the Creation: The Rio Earth Summit: Challenges for the Churches* (Geneva: World Council of Churches, 1992), 17. Despite the intuitive appeal of this story one needs to note that Britain's standard of living has increased dramatically since 1947, while the so-called consumer class in India has grown to around 150–200 million people, despite extreme levels of poverty that remain prevalent in that country.
12. The Sustainable Development Goals is a set of 17 global goals with 169 targets contained in paragraph 54 of United Nations Resolution A/RES/70/1 of September 25, 2015. It is officially referred to as "Transforming our World: the 2030 Agenda for Sustainable Development." It replaced the Millennium Development Goals as an agenda for development after 2015.
13. See *Our Common Future*, Chapter II, note 63.
14. For Timothy Morton, likewise, sustainable capitalism and military intelligence are further examples of such contradictions! See Morton, *Hyperobjects*, 111.

15. See Midge Béguin-Austin (ed.), *Sustainable Growth – Contradiction in Terms?* (Geneva: The Ecumenical Institute, 1993).

16. De Gruchy, *Keeping Body and Soul Together*, 52.

17. Ibid., 59.

18. Gonzalez, "Bridging the North-South Divide," 417. Gonzalez adds (with reference to a formulation by Gilbert Rist) that the growth policy promoted in the Brundtland report "that is supposed to reduce poverty and stabilize the ecosystem hardly differs at all from the policy which historically opened the gulf between rich and poor and placed the environment in danger" (419). The reference is to Gilbert Rist's *The History of Development: From Western Origins to Global Faith* (1997), 186.

19. See Leonardo Boff, *Cry of the Earth, Cry of the Poor* (Maryknoll: Orbis, 1997), 101.

20. Ibid., 67.

21. See Swilling and Annecke, *Just Transitions*, 23.

22. Bonneuil and Fressoz, *The Shock of the Anthropocene*, 22.

23. See my provocative question in an article entitled Ernst M. Conradie, "Why Can't the Term Development Just be Dropped Altogether? Some Reflections on the Concept of Maturation as Alternative to Development Discourse," *HTS Theological Studies* 72:4 (2016): 1–11. I noted with interest that Bill McKibben also employs the concept of maturation (growing up) instead of growth (growing). See his *Falter: Has the Human Game Begun to Play Itself Out?* (New York: Henry Holt and Company, 2019), 227.

24. Quoted in Rasmussen, *Earth Community, Earth Ethics*, 112–13.

25. See *An Ecomodernist Manifesto*, 10.

26. See Ibid., 31.

27. Sachs, *Common Wealth*, 206.

28. Ibid., 206–12.

29. SACC, *Climate Change*, 26–27.

30. Ibid., 27–28.

31. See Bonneuil and Fressoz, *The Shock of the Anthropocene*, 250.

32. See Paul Ekins and Michael Jacobs, "Environmental Sustainability and the Growth of GDP: Conditions for Compatibility," in *The North, the South and the Environment: Ecological Constraints and the Global Economy*, edited by Vinit Bhaskar and Andrew Glyn (London: Earthscan Publications, 1995), 9–46 (26).

33. A UNEP report entitled *Decoupling Natural Resource Use and Environmental Impacts from Economic Growth: A Report of the Working Group on Decoupling to the International Resource Panel* (2011) indicates that, on a worldwide scale, resource consumption is steeply on the rise while resource consumption is still a reliable companion of economic prosperity (ix). Development can only be sustainable if it is without growth in throughput of matter and energy beyond regenerative and absorptive capacities. This would require resource "decoupling," that is, "using less resources per unit of economic output and reducing the environmental impact of any resources that are used or economic activities that are undertaken" (xiii). Dematerialization then entails using less materials, energy, water, and land resources for the same economic output. See also Swilling and Annecke, *Just Transitions*, 72–80.

34. *An Ecomodernist Manifesto* adds a distinction between relative and absolute decoupling: "Decoupling occurs in both relative and absolute terms. Relative decoupling means that human environmental impacts rise at a slower rate than overall economic growth. Thus, for each unit of economic output, less environmental impact (e.g., deforestation, defaunation, pollution) results. Overall impacts may still increase, just at a slower rate than would otherwise be the case. Absolute decoupling occurs when total environmental impacts—impacts in the aggregate—peak and begin to decline, even as the economy continues to grow," 11.

35. See the roadmap for decarbonization sketched by Johan Rockström and colleagues, with targets of halving carbon emission for each decade, "A Roadmap for Rapid Decarbonisation," *Science* 355:6331 (2017): 1269–71.

36. Swilling and Annecke, *Just Transitions*, 99–100. They add that development economists may find it "impossible to reconcile non-material growth with poverty eradication," but insist that "the poor will be the biggest losers in an unsustainable world" (100). Moreover, developing economies will be forced to rethink strategies on economic growth if this means increasing biophysical throughput per capita to levels similar to so-called developed economies (103).

37. There may be some hope here as far as industrialized economies are concerned. The report on *Global Change and the Earth System* by Will Steffen and colleagues maintains, "In general, a trend of dematerialisation can be seen in many advanced economies . . . in which the value and amount of economic activity continue to grow but the amount of physical material flowing through the economy does not. This leads to a decoupling of economic growth and pollution, an evolving form of mitigation against global change" (290). The question remains whether such growth can be maintained with a significant reduction of biophysical throughput, given the need to drastically reduce reliance on fossil fuels in industrialized countries.

38. See https://www.globalcarbonproject.org/carbonbudget/19/highlights.htm (published December 4, 2019, accessed December 10, 2019). The figures given can be confusing for nonexperts, especially given the distinction between carbon, carbon dioxide (with added weight for the oxygen attached to the carbon), and carbon dioxide equivalent (including other greenhouse gases).

39. Lynas, *The God Species*, 214.

40. Ibid., 238–41.

41. Ibid., 241.

42. Ibid., 241–42.

43. Galtung, *Peace by Peaceful Means*, 223.

44. Ibid., 229.

45. See Goudzwaard, *Idols of Our Time*, 104.

46. Ibid., 105.

47. Steffen et al., "The Anthropocene: Conceptual and Historical Perspectives," 861–62.

48. Hamilton, *Requiem for a Species*, 38.

49. Ibid., 33.

50. Ibid., 64.

51. Hamilton, *Earthmasters*, 207.

52. Yuval Harari, *Homo Deus: A Brief History of Tomorrow* (London: Vintage, 2017), 23.

53. Ibid., 244.

54. Klein, *This Changes Everything*, 252.

55. Ibid., 21.

56. Angus, *Facing the Anthropocene*, 112.

57. Ibid., 113.

58. See Lewis and Maslin, *The Human Planet*, 383–84.

59. See Steffen et al., *Global Change and the Earth System*, 286.

60. See http://w2.vatican.va/content/francesco/en/apost_exhortations/documents/papa-francesco_esortazione-ap_20131124_evangelii-gaudium.html (accessed January 12, 2019).

61. The emphasis on a recognition of limits is striking in §204: "The current global situation engenders a feeling of instability and uncertainty, which in turn becomes 'a seedbed for collective selfishness.' When people become self-centred and self-enclosed, their greed increases. The emptier a person's heart is, the more he or she needs things to buy, own and consume. It becomes almost impossible to accept the limits imposed by reality. In this horizon, a genuine sense of the common good also disappears. As these attitudes become more widespread, social norms are respected only to the extent that they do not clash with personal needs. So our concern cannot be limited merely to the threat of extreme weather events, but must also extend to the catastrophic consequences of social unrest. Obsession with a consumerist lifestyle, above all when few people are capable of maintaining it, can only lead to violence and mutual destruction."

62. See Yusoff, *A Billion Black Anthropocenes or None*, 51. Following Achille Mbembe, she adds that blackness is established as an exception to this coda.

63. Virginie Maris, "Back to the Holocene: A Conceptual, and Possibly Practical, Return to a Nature not Intended for Humans," in *The Anthropocene and the Global Environmental Crisis*, edited by Clive Hamilton, Christophe Bonneuil and François Gemene (London and New York: Routledge, 2015), 123–33 (130–31).

64. See Wood, *Deep Time, Dark Times*, 26.

65. See Serres, *The Natural Contract*. Serres proposes the need for a "natural contract" that extends the social contract since the world is typically absent from the latter. Indeed, "We must decide on peace among ourselves to protect the world, and peace with the world to protect ourselves" (25). For Serres, the natural contract is one of symbiosis and reciprocity: "Our relationship to things would set aside mastery and possession in favor of admiring attention, reciprocity, contemplation and respect; where knowledge would no longer imply property, nor action mastery, nor would property and mastery imply their excremental results and origins" (38). The natural contract implies for him the metaphysical recognition that collective decisions operate and is embedded in the same natural world as all others. The natural contract acknowledges an equilibrium between human power and the forces of the world, just as a social contract recognizes equality between human signatories (45–46). Such a contract is necessary in a world where science prevails over law (61).

66. See the excellent discussion on "Creation, Nature, and Dualism" by Hans Dirk van Hoogstraaten, *Deep Economy: Caring for Ecology, Humanity and Religion* (Cambridge: James Clarke & Co., 2001), 61–84 (78).

67. See also, remarkably, Steffen et al., *Global Change and the Earth System*, 247.

68. See the comment by Michel Serres: "A parasite—which is what we are now—condemns to death the one he pillages and inhabits, not realizing that in the long run he's condemning himself to death too." See *The Natural Contract*, 38.

69. Ibid., 33.

70. Lynn White, "The Historical Roots of our Ecological Crisis," *Science* 155 (1967): 1203–7 (1205).

71. Bruno Latour, following Peter Sloterdijk, points to an oddity in terminology since the Copernican "revolution." If one adopts a theocentric approach, then the Earth as planet becomes peripheral and revolves around God. If the Earth is in the center, then God is relegated to the periphery. This problem dissipates in a panentheist position where God encompasses the whole history of the universe, but then the metaphor "center" in a theo*centric* approach does not hold. A Trinitarian theology may still hold that Christ is the hinge upon which history turns (thus Christocentric) but that Creator and Spirit moves around that center. See Latour, *Facing Gaia*, 125.

72. Harari, *Homo Deus*, 114–15.

73. Ibid., 76.

74. With Donna Haraway, I warm to the suggestion that we human beings are best understood not as being posthuman but as being compost (humus). Given the toxins and metals in human bodies, other organic material may be better than humans at being compost though. See Haraway, *Staying with the Trouble*, 101.

75. See Neiman, *Evil in Modern Thought*, 228, in conversation with Freud. She observes that the old metaphor of being at home expresses some reduced expectations—no longer hoping for towers that would reach up to the heavens or a great chain of being that bears witness to God's glory. Instead, she says, "We merely seek assurance that we find ourselves in a place where we know our way about" (304). Despite such modesty, such assurance is not readily available in a disenchanted world where the argument from design has been lost, where natural evil abounds and where social evil undermines any possibility of being at home, even in one's own house. She adds, "Decency demands that we refuse to feel at home in any particular structure the world provides to domesticate us" (305). Modern consciousness (after Lisbon) required from us "to stop viewing the world as the home that a stern but indulgent parent might have built us, and to grow up and build our own" (305). Auschwitz revealed that house to be a concentration camp.

76. See Latour, *Facing Gaia*, 14.

77. Pinker, *Enlightenment Now*, 134, 136. It must be noted that Pinker's concern is not with the Anthropocene. Although he discusses the challenges associated with climate change (from an ecomodernist perspective), the term Anthropocene does not appear in the index of *Enlightenment Now*.

78. Ibid., 452–53.

79. Michel Serres refers to this becoming as hominescence. See his *Hominescence* (London: Bloomsbury, 2019).

80. David Wallace-Wells, *The Uninhabitable Earth: Life after Warming* (New York: The Duggan Books, 2019), 175.

81. Harari, *Homo Deus*, 408.

82. Ibid., 444.

83. Ibid., 453–54.

84. Ibid., 458.

85. See Serres, *Hominescence*, 62.

86. Subsequent references in parenthesis in the text are to Hamilton's *Defiant Earth*.

87. Hamilton recognizes the theological implications of the ecomodernist Nietzschean claim to kill God and to replace God by humans: "If in theology man is the creation of God in his image, and therefore always subject to a greater power, in the ecomoderns' anthropology man is the creature of nature as its highest living form. This nature is a not a power that rules over man; the tables are turned, and man rules over nature" (67). Human-directed progress usurps the place of God as a secularized notion of divine providence (70). Hamilton's reading of the notion of providence (67–71) is provocative and deserves more detailed discussion than what can be offered here.

88. See *An Ecomodernist Manifesto*, 31.

89. With reference to Ewa Domanska, "Beyond Anthropocentrism in Historical Studies," *Historien* 10 (2010): 118–30 (118).

90. This is indicative of Hamilton's position:

> "Self-organizing systems always have a tendency; they act as if they have an intended result. And is it not reasonable to think that the drive that saw humans spread across the Earth and increasingly transform the environment could at some stage become so large and active that it would rival some of the great forces of nature? It was not inevitable, for there were countervailing forces restraining thoughtless exploitation of nature, from the Romantics to the modern environmental movement in all its variations. Yet the propensity to end up where we are, was strong enough to give rise to a legitimate sense of directionality in the history of human-Earth relations" (122).

91. See Latour, *Facing Gaia*, 117.

92. Chakrabarty, "Postcolonial Studies and the Challenge of Climate Change," 1–2.

93. Ibid., 2.

94. Ibid., 9.

95. Ibid.

96. Ibid., 14.

97. Ibid., 14–15.

98. Chakrabarty, "The Climate of History," 220.

99. See Dipesh Chakrabrty, "The Anthropocene and the Convergence of Histories," in *The Anthropocene and the Global Environmental Crisis*, edited by Clive Hamilton, Christophe Bonneuil, and François Gemene (London and New York: Routledge, 2015), 44–56 (53).

100. See Latour, *Facing Gaia*, 126.

101. Quoted in Anthony Lowes, "Up Close and Personal: In the End, Matter Matters," in *Earth Revealing – Earth Healing: Ecology and Christian Theology*, edited by Denis Edwards (Collegeville: The Liturgical Press, 2001), 125–44 (129). The reference is to Kelly's *An Expanding Theology* (Newtown: E.J. Dwyer, 1993), 133. This has theological implications for language about God. Postmodern suspicions have rightly undermined the liberal (and deeply anthropocentric) confidence that our understanding of God can be based on an understanding of the self.

102. See Holmes Rolston (III), *Conserving Natural Value* (New York: Columbia University Press, 1994), 158–62. This insight helps to avoid the, in my view, flawed distinction between a "weak" and a "strong" form of anthropocentrism.

103. Holmes Rolston (III), *Environmental Ethics: Duties to and Values in the Natural World* (Philadelphia: Temple University Press, 1988), 112.

104. See Callicot, *Thinking Like a Planet*, 9–11.

105. Any theological legitimation of such anthropocentrism has therefore been heavily criticized in Christian ecotheology. In the words of James Nash, "The traditional idea that the earth, or even the universe, was created solely for humans is, in our scientific age, sinfully arrogant, biologically naïve, cosmologically silly and therefore theologically indefensible." See James A. Nash, "Towards the Ecological Reformation of Christianity," *Interpretation* 50:1 (1996): 5–15 (8).

106. See Serres, *The Natural Contract*, 33.

107. See Hans Dirk van Hoogstraaten, *Geld en Geest: Over Milieu-ethiek* (Baarn: Ten Have, 1993), 63.

108. See, Serres, *Malfesance*, 85.

109. See Lewis and Maslin, *The Human Planet*, 16.

110. Michael Northcott, "On Going Gently into the Anthropocene," in *Religion in the Anthropocene*, edited by Celia E. Deane-Drummond, Sigurd Bergmann, and Markus Vogt (Eugene: Cascade Books, 2017), 19–34 (24).

111. Sallie McFague, *The Body of God: An Ecological Theology* (Minneapolis: Fortress Press, 1993), 109.

112. See, Serres, *Malfeasance*, 54.

113. Gerrit Manenschijn, *Geplunderde Aarde, Getergde Hemel: Ontwerp voor een Christelijke Milieu-ethiek* (Baarn: Ten Have, 1988), 126–27.

114. Philip Hefner, *The Human Factor: Evolution, Culture and Religion* (Minneapolis: Fortress Press, 1983), 49.

115. Most theories of human uniqueness based on tool use, tool making, culture, food sharing, theory of mind, planning, empathy, morality, or inferential reasoning have collapsed since such features have been observed among wild primates. See Frans de Waal's whole oeuvre and, quite explicitly, his article "Who Apes Whom," http://www.nytimes.com/2015/09/15/opinion/who-apes-whom.html? (accessed November 6, 2016). The one exception, De Waal admits, is human language capacity based on the use of symbols. In this regard, many follow Terrence W. Deacon, *The Symbolic Species: The Co-evolution of Language and the Brain* (New York: W.W. Norton, 1997). Deacon distinguishes between three forms of signs, namely, icons, indices, and symbols. With icons there is a similarity between the sign and an object (e.g., drawings), indices are mediated by some physical or temporal connection

between sign and object (e.g., a thermometer), while there is some agreed-upon association between a symbol and its reference, irrespective of the physical characteristics of the sign and its referent (e.g., a national flag) (see 70). Only humans display symbolic behavior. In Christian theology, see also J. Wentzel van Huyssteen, *Alone in the World: Human Uniqueness in Science and Theology* (Grand Rapids: Eerdmans, 2006). See also Van Huyssteen's essay, "Questions, Challenges, and Concerns for the Image of God," in *Finding Ourselves after Darwin*, edited by Stanley P. Rosenberg (Grand Rapids: Baker Academic, 2018), 33–47. He captures the emerging consensus: "In the Upper Paleolithic in Europe, and probably earlier in Africa, anatomically modern humans crossed a new cognitive threshold into a semiotic realm, a threshold of semiotic competence that allowed for the combination of remarkable new forms of symbolic communication" (41).

116. See Serres, *The Incandescent*, 212.

117. See Kolbert, *The Sixth Extinction*, 258.

118. See Serres, *The Incandescent*, 109–10. In his evocative language Serres hopes that a Grand Narrative can be reconstructed that will allow for a humanism to emerge that is finally worthy of its name, a humanism that is "single and universal since written in the encyclopedic language of all the sciences and since it can be translated into every vernacular without particularism or imperialism" (231). This Grand Narrative follows the journey of cosmic, biological, and hominid evolution; the common origin of humanity; the spread of humans across the globe; and the contemporary coming together of a diversity of languages and cultures.

119. Sloterdijk, *In the World Interior of Capital*, 148.

120. Immanuel Kant, *The Critique of Practical Reason* 6:162, as quoted and discussed in Paul Guyer, "Introduction: The Starry Heavens and the Moral Law," in *The Cambridge Companion to Kant and Modern Philosophy*, edited by Paul Guyer (Cambridge: Cambridge University Press, 2006), 1–27.

121. Serres, *The Incandescent*. Remarkably, Serres links this glowing with Pentecost: "What if we live and thought in the flames, the wind, the breath, the incandescent tongues of fire" (187).

122. Serres, *The Natural Contract*, 95.

123. See Neiman, *Evil in Modern Thought*, 92.

124. See again my "Penultimate Perspectives on the Root Causes of Environmental Destruction in Africa," also *Redeeming Sin?*, Chapter 2.

125. See Chakrabarty, "The Politics of Climate Change," 31.

126. Subsequent references in parenthesis in the text below are to Hamilton's *Defiant Earth*.

127. Latour illustrates this point with a literary reference to the creation of a monster by Dr. Frankenstein as a cautionary tale about the dangers of technology.

> "Dr. Frankenstein's crime was not that he invented a creature through some combination of hubris and high technology, but rather that he abandoned the creature to itself. . . . Written at the dawn of the great technological revolutions that would define the 19th and 20th centuries, Frankenstein foresees that the gigantic sins that were to be committed would hide a much greater sin. It is not the case that we have failed to care for Creation, but that we have failed to care for our technological creations. We confuse the monster

for its creator and blame our sins against Nature upon our creations. But our sin is not that we created technologies but that we failed to love and care for them. It is as if we decided that we were unable to follow through with the education of our children." The sins of sloth seem to outweigh the sins of hubris here! See Bruno Latour, "Love Your Monsters."

128. See Mbembe, "Decolonizing Knowledge and the Question of the Archive," my emphasis.

129. Quoted in Swilling and Annecke, *Just Transitions*, 25. The reference is to a brief article, "Power Shift," *Resurgence*, August 2002, 32–34.

Chapter 6

Quasi-Soteriologies in the Anthropocene?

APARTHEID AS QUASI-SOTERIOLOGY

Apartheid in South Africa functioned not only as an ideology but in the theological critique of apartheid it was also described as a pseudo-gospel. This suggests that people also "believed in" apartheid. Accordingly, some people put their trust in the system, assuming that this was the best way of structuring society, that this was the only option to regulate relations between different races, and that this was the only way to ensure well-being for themselves and indeed for all others. It was assumed that the only way in which peace between different racial groups could be maintained was by separating them from one another as far as this may be possible. A clear expression of this critique may be found in an essay by Jaap Durand, the former vice-rector of the University of the Western Cape:

> Apartheid in South Africa is much more than a political system imposed on South African society through many years of legislation. It is a comprehensive ideology and view of life involving the organization and control of man and society with the pretence of a pseudo-gospel: the solution of social problems through—in this instance—the legal separation of groups and individuals on the basis of race and colour. I speak deliberately of a "pseudo-gospel" because the "solution" that is offered is given biblical sanction. Apartheid is more than a political attempt to surmount the problems of a complicated society: it is God's will for southern Africa, expressed in his creative order and opposed only at the risk of dire consequences.[1]

The conviction that the differences between racial groups are more fundamental than what binds them together is the key to understanding the way in which apartheid was elevated toward a quasi-soteriology. Gradually it led to the intuitive assumption that separation on the basis of race provides the only way toward well-being and prosperity for people of mainly European descent. This was regarded as the only strategy that could work in the long term.

This way in which separation was assumed to be a precondition for salvation could be identified at different levels and in different spheres of society. In an earlier contribution, Murray Coetzee and I argued that such a quasi-soteriology had political, military, economic, social, cultural, and ecclesial dimensions. Let me draw on this contribution to describe only the cultural assumptions that reinforced such a quasi-soteriology:[2]

Social segregation in South Africa was based on a deep sense of cultural superiority. This was associated with Western civilization, best expressed by the cultural refinement of the Afrikaner elite, including pastors, teachers, and cultural leaders. This notion of "civilization" was associated with technological advances, superior weapons, literacy, hygiene, dress code, culinary habits, medical services, transport, higher education, and an appreciation of "higher" forms of culture such as literature, poetry, music, and the arts. This sense of superiority was reinforced by class differences and by separation that prevented ongoing exposure to the language, customs, worldviews, and religious beliefs of others. Here apartheid was regarded as the only way to protect "Western civilized standards" against the surrounding "sea of barbarism." The disgust for the way of life of others, spurred on by a deeply embedded racism, may be regarded as the deepest reason why people believed in apartheid, why it became so deeply entrenched in the hearts and minds of ordinary people, and why it was far more than just a political system. In order to maintain a particular cultural identity and a sense of superiority, clear distinctions had to be drawn between civilization and savagery. The notorious laws on mixed marriages and immorality (sexual intercourse across racial lines) had to reinforce such distinctions, to protect Afrikaners against themselves in order to maintain their identity and hence their power.

Moreover, this sense of cultural identity and superiority had to be protected against the "swamp of barbarism" in the midst of Afrikaners, not only the "sea of barbarism" from the outside. This has to be understood against the background of the living conditions associated with "poor whites" in urban areas, especially after the South African War (1899–1902). The task of cultural leaders, women's groups, the press, opinion formers (including the Afrikaner Broederbond), and pastors alike was to "uplift" such people—to provide education and training, to promote job creation, to help ensure better living conditions, and to portray, embody, and practice the ideals of "Western civilization." The differences between such white "civilized standards" and the suction power of the surrounding traps of poverty had to be drawn as clearly as possible in order to encourage social upliftment among poor whites. The elite in every town had to set and raise the standards of cultural refinement. Perhaps the need for emphasizing such a sense of cultural superiority was born from feeling threatened or from an even deeper sense of inferiority, which probably resulted from decades of cultural and linguistic imperialism under British rule.

These observations also illustrate the continuing lure of apartheid thinking, in South Africa and elsewhere in the world. One may argue that such elitism is indeed very widespread. What made apartheid relatively crude was that such a need for distinction was directly associated with race classification. This was an all too simplified way of drawing such a distinction, but was nevertheless enforced stubbornly and foolishly. Even when it became evident that the political policies associated with segregation, apartheid, "separate development," or "plural relations" could no longer work, no viable alternative could be conceived. The trust in a system of separating people on the basis of race was too deeply entrenched. The obstinacy that characterized the fierce suppression of resistance against this system was indicative of this. Apartheid was indeed regarded as the only available means of "salvation."

QUASI-SOTERIOLOGY IN THE ANTHROPOCENE?

How may this analysis be relevant in the Anthropocene? In industrialized societies, the Christian message of salvation (perceived to be aimed at securing eternal life with God in heaven for an immortal soul) has typically been discarded through a focus on the here and now. This is exacerbated by the way in which salvation is understood in consumer societies. Jay McDaniel comments that such salvation comes on the basis of appearance, affluence, and marketable achievement:

> We feel saved, or made whole, when we appear (or think we appear) to be successful and attractive to others, when we feel good about how much money we have (which always has to be as much as, or more than, others), and when we can take pride in accomplishments that make sense in the marketplace. The "saints" of consumerism are, then, the people who are "fully saved" by these standards. They are the successful executives of very large corporations and the celebrities who "have it all" and who show us what it is like to have it all.[3]

Such a focus on the here and now also allows for an emphasis on liberation from oppression, the emancipation of women, decolonization in response to colonial conquest, peace-making amid violent conflicts, and ecological healing/restoration. However, in industrialized societies with stable democracies, capital investments,[4] education, health care, labor-saving devices, insurance, pension funds, and the welfare state have been operating as functional replacements for the promise of salvation. It is from the invisible hand of the market, regulated by the state, that one may expect employment, housing, and economic security.[5] Even though such invisibility operates more like a black hole and a black box (Michel Serres's term for the open pit of slavery[6])! Where these are not enough, social security, safety nets, and therapy are solicited to attend to human well-being. Where these still do not suffice, utopian political movements provide a materialized form of the Christian promise of salvation.[7] Where revolution does not deliver on its promises, one may still find solace in sex, drugs, alcohol, or religion as the opium of the people.

The Anthropocene brings the recognition that none of these would suffice to "appease" the "awakened monster." There are no quick-fix solutions or god-tricks intervening from above that would do the trick. We should not place our hope in a people's army replacing the state, a revolution in consciousness, or even in eschatological solutions from the sky (a meteorite?).[8] Neither would faith in Growth, History, Knowledge, Modernity, Nature, Progress, Providence, Reason, Science, or the State help to save us. As Carmen Gonzalez has it, the "quasi-religious belief in the benefits of liberalized trade has produced an international economic order that generally ignores the environmental and social consequences of production and implicitly encourages environmental subsidies."[9] And as Donna Haraway observes, those who believe that they have the answers are terribly dangerous.[10]

What about technology, the wishful thinking that something will come along, the trust that human ingenuity will somehow offer solutions to mitigate carbon emissions and thus curb climate change? "Should anything save us," declares journalist David Wallace-Wells, "it will be technology."[11] It is evident that many do place their trust here. The "Ecomodernist Manifesto" is a pertinent example in its call "to use humanity's extraordinary powers in service of creating a good Anthropocene,"[12] one that would promote the four human aspirations of peace, freedom, material well-being, and environmental health together:[13]

> Meaningful climate mitigation is fundamentally a technological challenge. By this we mean that even dramatic limits to per capita global consumption would be insufficient to achieve significant climate mitigation. Absent profound technological change there is no credible path to meaningful climate mitigation. While advocates differ in the particular mix of technologies they favor, we are aware of no quantified climate mitigation scenario in which technological change is not responsible for the vast majority of emissions cuts.[14]

It seems obvious that climate change cannot be addressed without appropriate forms of technology. Yet, one may be forgiven to wonder whether the instruments that have exacerbated the underlying problem can really be of service to remedy the problem. To put this more precisely: It is not the "monster" of technology that is to be blamed, or indeed the ingenuity that facilitated the development of such technology? It is the unwise use of such technology by particular human societies that led to the Anthropocene. It would not help to adopt more sophisticated technology if that underlying problem is not addressed.

The problem is that these very sources of salvation (and their carbon footprint) are also the root causes of the destruction associated with a changing climate, the loss of biodiversity, and ocean acidification. In addition to the oppressive past that many seek to escape from, there is now a daunting future that we would want to avoid so that any new proposal for an emancipatory project must transcend the progressive categories of the past.[15] For Clive Hamilton any idea of "saving the planet" is now obtuse. The planet is no longer the Holocene one that we may be trying to save, but something else that we may not wish to "save": "Gaia [is] not the all-loving, all-nurturing Mother Earth of the romantics but more like the half-crazed, bloodthirsty and vindictive goddess of the original Greek tales." Nature is no longer only a becalming Mother but also a terrifying Other. Mother Nature opens her arms not to embrace but to crush us.[16] This has soteriological implications: "Our goal can no longer be to 'save nature' but to save ourselves, from ourselves and from nature, knowing that every disturbance to the Earth System reduces the chances of doing so."[17] This suggests a remarkable reversal of the core insight in Christian ecotheology that salvation has to be understood not as salvation from the Earth but as the salvation of the whole Earth.[18] We have awakened the slumbering beast and the Promethean hope to anaesthetize it with geoengineering is "foolishness on an epic scale."[19] Hamilton finds himself unable to invest faith in divine providence: "Where once we could fear and love God and truly believe in him and his saving power, now we can only fear Gaia. But Gaia is no messiah, which leaves self-preservation as the only motive, a negative motive that seems much too weak."[20] The gods have absconded long ago, although new strange gods may be awaiting us beyond the horizon. We do not know our fate but must learn to live on this planet in the half-light of not knowing.[21] Moreover, any attempt to declare the Anthropocene "good" is structurally similar to a theodicy, that is, the theological argument that aims to prove the ultimate benevolence of God despite the overwhelming presence of evil.[22] This, he argues, can only serve the purpose of condoning suffering and oppression. This also applies to the term "good Anthropocene": "For the victims inclined to protest against the system, the golden promise of a new dawn lulls them into silent endurance.

The message of the good Anthropocene to those suffering now and in the future from human-induced droughts, floods and heat waves is: You are suffering for the greater good; we will help to alleviate it if we can but your pain is justified."[23]

If the diagnosis of the underlying problem is located in Earth systems, then the remedy presumably has to be found there too. This poses an intriguing dilemma. On the one hand, the human power to change Earth systems is clearly evident. Many have therefore suggested that "the end of nature" (Bill McKibben) or even the "death of nature" (Carolyn Merchant) has arrived.[24] While with Noah's flood culture disappeared beneath nature, with the current deluge nature disappears beneath culture.[25] On the other hand, there is the recognition that Earth systems can be influenced but cannot be controlled by humans, simply because of the scale involved. Once the "monster" is awakened, it will be impossible to control or even appease; nature returns with a vengeance. The announcement of the end of nature is then premature; it may be appropriate to Holocene conditions, but not to the Anthropocene.[26] What, then, can save human civilization from collapse toward the end of the twenty-first century? Should we look for a Messiah (or Führer?) to deliver us from evil, knowing that any messianic claims may well prove to be false? Hamilton confronts the central question confronting humankind: "What can be done to slow the changes; what can be done to adapt to those that are unavoidable; and, over the long term, what can be done to remediate the damage to the Earth System?"[27] The one obvious, widely discussed candidate for such salvation is geoengineering.

GEOENGINEERING AS QUASI-SOTERIOLOGY?

Geoengineering is defined by the IPCC as "a broad set of methods and technologies that aim to deliberately alter the climate system in order to alleviate the impacts of climate change." It adds:

> Most, but not all, methods seek to either (1) reduce the amount of absorbed solar energy in the climate system (Solar Radiation Management) or (2) increase net carbon sinks from the atmosphere at a scale sufficiently large to alter the climate (Carbon Dioxide Removal). Scale and intent are of central importance. Two key characteristics of geoengineering methods of particular concern are that they use or affect the climate system (e.g., atmosphere, land or ocean) globally or regionally and/or could have substantive unintended effects that cross national boundaries. Geoengineering is different from weather modification and ecological engineering, but the boundary can be fuzzy.[28]

Indeed geoengineering comes in many different forms, including altering farming practices on a large scale (toward less meat-intensive diets), reforestation, carbon sequestration and fine-tuning the atmosphere's thermostat to avoid both the onset of new ice ages and overheating through remedial steps (solar radiation management).[29] While carbon sequestration responds to the causes of the malady (too much carbon in the atmosphere) and the failure of mitigation efforts, solar radiation management responds to the symptoms associated with too much heat and the expected failure of adaptation efforts.

The technical feasibility, affordability, mitigating impact, and timeous implementation of such plans for geoengineering are under continuous public scrutiny. Jeffrey Sachs, for example, with the backing of colleagues at the Earth Institute at Columbia University, affirms that the technology for carbon capturing is already available, that technologies can be enhanced for carbon sequestration (the storage or disposal of carbon dioxide, e.g., in abandoned oil wells, beneath the ocean floor, or in underground saline aquifers) at a large scale, and that the costs for this need not be too prohibitive. He is clearly attracted to such technologies since the economic costs of scaling down on the use of fossil fuels may choke economic development, especially in poor economies. He admits that the scale of such an operation still poses problems for feasibility but then suggests that more funding for research and development be granted for capturing carbon dioxide directly from the air at suitable locations.[30] The problem remains how such carbon capturing can be done at scale and what to do with the carbon once removed which will require massive pipelines that run the current fossil fuel industry in reverse.[31]

Erle Ellis acknowledges that geoengineering is the most "Promethean" form of environmental governance. He is fully aware of the controversies in this regard given the dangers of arrogance and recklessness, but articulates what constitutes the bottom line for its advocates:

> Interest in climate geoengineering has never been greater. The more Earth warms, the more it will cost to deal with the consequences, including disrupted food systems, extended droughts, extreme heat waves, rising seas, severe storms, and other harm to societies. To date, societal efforts are failing to halt rising greenhouse gas emissions. The less that is done now, the more societies will be willing to do for a cooler climate in the future. And geoengineering may turn out to be the surest way to achieve this.[32]

It would be facetious to reject any form of geoengineering outright as being too risky since it may also be too risky not to attend to possibilities in this regard. Such assessments typically refer to a widely cited comment by Paul Crutzen made in 2007: "The very best would be if emissions of the greenhouse gases could be reduced so much that the stratospheric sulfur release experiment would not need to take place. Currently, this looks like a pious wish."[33] This danger of not considering options for geoengineering is also recognized by Janos Pasztor when he states, "Sooner rather than later, policymakers around the world will need to confront an uncomfortable reality: that despite the best efforts of national governments and thousands of mayors and other civic leaders, we can no longer contain global average temperatures to below 1.5–2 degrees Celsius above preindustrial levels through mitigation of carbon emissions alone."[34] He adds that there are risks in *not* using a technology that could provide many potential benefits and help mitigate the worst effects of climate change.[35] A purist position is hardly tenable to the extent that the Anthropocene itself is already the product of massive geoengineering processes. Admittedly, if intentional altering of the atmosphere is inscribed in the definition of geoengineering, this statement could be questioned, but then carbon mitigation is also intentional and would affect the atmosphere too.

The bottom line, as Mark Lynas has it, is that flawed solar radiation management with serious unintended outcomes may still be better for planetary well-being than five or six degrees of global warming. Inadvertent forms of geoengineering may well be worse than deliberately planned forms.[36] It is as if his own book entitled *Six Degrees*, and the failure of mitigation efforts in the ten years since that book was published, forced him to opt for the proliferation of nuclear energy and to consider the necessity of geoengineering. While others warn against the danger of "playing God,"[37] Lynas insists that "playing God is good for the planet."[38] While some are concerned about prideful action, others are concerned with slothful inaction![39]

In David Grinspoon's assessment, "on its own, Earth is capricious and cruel. If we want to keep our civilization for long enough, we're going to have to deal with that. On the longer timescales of ice ages, in order to prevent these deathly oscillations, we will have to learn to provide negative feedback that is missing from the natural Earth system."[40] Grinspoon recognizes the dangers of human arrogance. He puts the following words in Carl Sagan's mouth: "It would be an act of consummate recklessness and arrogance to try to jury-rig the only home we have, to attempt a quick-and-dirty repair job on the complex, deeply mysterious, and exquisitely balanced global mechanisms that all human life, and all other life we know of, depends upon, when we are still so ignorant about their functioning."[41] If so, "we really have no choice but to control our CO_2 habit."[42] It seems, then, that we as human beings are free to set the global thermostat wherever it pleases us, if we are able to do so. Either way, it is not as if we can avoid changing the Earth's temperature since we are doing so already. The question is whether we can adjust the atmosphere like a giant air conditioner to align it to our economic and cultural rhythms[43] (especially those who have air conditioning at home and at work).

In *Love Your Monsters* Shellenberger and Nordhaus regard technological risk-taking as an unavoidable, indeed necessary part of the human condition. It is savior technologies that will save us from the worst effects of climate change.[44] They refuse to depart from the ideology of dominating nature or to heed the precautionary principle but insist on the following:

> Each new act of salvation will result in new unintended consequences, which ill in turn require new acts of salvation. What we call "saving the earth" will, in practice, require creating and recreating it again and again for as long as humans inhabit it.[45]

Steven Pinker advocates a more restrained use of geoengineering since he recognizes the "moral hazard" that would tempt countries to shirk their responsibility for reducing greenhouse gas emissions. He suggests forms of solar radiation management that is moderate (just enough to reduce the rate of warming), responsive (gradual, constantly monitored, and adjusted), and temporary (that would allow some breathing space to eliminate greenhouse gas emissions and to develop carbon sequestration technologies). He remains optimistic that such technologies will become available within the next fifty years or so although his optimism is not a complacent one and requires a retrieval of reason, science, humanism, and progress.[46]

But what if such strategies do not work either, and if the unintended outcomes are simply added to the impact of global warming? Is a "cascade of unanticipated consequences" not well-nigh inevitable given our lack of understanding the complexities of the Earth system?[47] How can you engineer something that you do not really understand? Is the danger of moral corruption not high here—in the sense that the consumer class can maintain their current standards of living in the hope that a green revolution will somehow be forthcoming?[48] What is the status of the argument that geoengineering may be the lesser of two evils?[49] Can one evil be distinguished from another so discretely? Can doing the lesser of two evils really be offered as a form of salvation?

The complex issues in this regard need not be reviewed here as there is by now ample available literature on the ethical issues involved. Let me mention only a few critical assessments that recognize the emergence of a quasi-soteriology:

The landmark report entitled *Global Change and the Earth System* (2004) stresses the precautionary principle on the basis of the growing realization that the Earth's natural variability is characterized by abrupt changes with catastrophic consequences for human societies. The precautionary principle helps to steer us away from any such disastrous pathways. The analogy adopted is that of a boat approaching a fatal drop over a waterfall. The boat's captain must make a decision well before becoming fully aware of the dangers that lie ahead. Human societies may already be in precisely this situation with their unintended experiment with the Earth system. Will Steffen and colleagues warn that

> it is exceedingly dangerous to apply simplistic technological solutions based on cause-effect logic within the context of complex, non-linear systems that are, as yet, poorly understood. In addition, there are serious ethical and moral considerations that will need to be debated and overcome across a very wide range of cultures and belief systems before any sort of global geo-engineering fix can be attempted.[50]

In an authoritative article, Will Steffen, Jacques Grinevald, Paul Crutzen, and John McNeill discuss the pros (offsetting a fraction of the anthropogenic increase in greenhouse gas warming) and cons (ocean acidification and altering precipitation patterns) of releasing aerosols in the stratosphere. Although it was Paul Crutzen who first raised the possibility of geoengineering solutions,[51] they offer a cautionary note:

> There is no doubt that, if geo-engineering is to play a significant role in preventing the climate system to warm beyond the "2°C guardrail," much more scientific research is required. Even more importantly, legal, ethical and societal issues, not to mention the challenges of global governance described earlier, will need to be thoroughly explored and solved before deliberate human modification of the climate system can be undertaken. Building trust among international political leaders of many different cultures and perspectives, and with the general public, would be required to make any large-scale climate modification acceptable, even if it would appear scientifically advantageous. Ultimately, the near inevitability of unforeseen consequences should give humanity pause for serious reflection before embarking on any geo-engineering approaches.[52]

In *Learning to Die in the Anthropocene* Roy Scranton concludes that "the problem is that global decarbonization is effectively incompatible with global capitalism. Capitalism needs to produce profit in order to spur investment. Profit requires growth. Global economic growth, even basic economic stability, depends on cheap, efficient energy."[53] On options for carbon capture and sequestration (CCS) he states:

> Absent government incentives and regulation, there's no reason for anyone to expect that carbon capture and sequestration will be developed or implemented on a meaningful scale, and as long as coal and oil companies remain politically powerful, there's no reason to expect robust enough government action in time for CCS to be a viable response to global warming.[54]

In *Earthmasters* Clive Hamilton observes that such Promethean plans have met with considerable resistance from those who distrust technological overreach. He adds the warning: "Nemesis waits in the shadows to punish Hubris. If Prometheus is the god of technological mastery, who is the Greek divinity of caution? Perhaps the closest is Soteria, the goddess of safety, reservation and deliverance from harm."[55] The soteriological vocabulary that Hamilton employs is noteworthy. In Greek mythology both Zeus and Dionysius carried the title "Soter" (as male counterpart). The Christian message of salvation is indeed far from unique. The question is not whether such salvation is needed, but who the savior may be, how that is to be done, and what salvation entails. The last of these depends on the diagnosis of what is wrong.

Yet, the danger is that desperate times will not only call for desperate measures but would also generate wannabe Messiahs (philanthropists, scientific experts, or think tanks) who may act unilaterally to "save the world," claiming that "there is no other alternative" to climate engineering to avert an emergency.[56] Even talking about climate engineering may undermine mitigation efforts. Using public funding for research, calling for more such funding, and testing such technologies may set the course for their ultimate deployment.[57] The "powerful predilection for finding excuses not to cut greenhouse emissions" creates a moral hazard.[58] In *Defiant Earth*, Hamilton comments on the "bizarre situation in which we have become potent enough to change the course of the Earth yet seem unable to regulate ourselves."[59]

In *This Changes Everything*, Naomi Klein resists the hope for a miracle cure, the belief that technology is going to save us from the effects of technology:

> Yet our faith in techno wizardry persists, embedded inside the superhero narrative that at the very last minute our best and brightest are going to save us from disaster. This is the great promise of geoengineering and it remains our culture's most powerful form of magical thinking.[60]

In *Staying with the Trouble* Donna Haraway grows impatient with the secular/religious "comic faith" in technofixes as a response to the Anthropocene/Capitalocene: "technology will somehow come to the rescue of its naughty but very clever children, or what amounts to the same thing, God will come to the rescue of his disobedient but very hopeful children."[61]

Or, as David Wood has it,

> Faith in technological fixes may blind us to the need to make changes of a different order. . . . if we put our faith in technology to square the circle, we would have to wait a long time. We do not have a long time.[62]

Upon further reflection, the hope is not that it is such technologies that will save us; the trust is in human ingenuity to come up with creative answers for new problems as they emerge. This comes with considerable financial backing from donors (with the hope of making money out of climate engineering?) and, of course, also interest from policy makers. However, such answers do not include any drastic political or economic interventions; the hope is to find "miracle" solutions, not to transform the hearts of minds of people. Indeed, as Jonathan Symons admits, "solar geoengineering is attractive precisely because it can be implemented without necessitating deep transformations within every community globally."[63]

The danger is that this logic offers three excuses to do nothing in the hope of finding technological solutions to solve social problems, namely, that it would buy time for a better response, that it would address an emergency situation, and that it would be more cost effective. Many observers fear that this poses a moral hazard: this can only undermine global carbon mitigation efforts and corrupt a sense of moral responsibility.[64] Arguments on the basis of declaring an emergency also need to be regarded with circumspection given the way oppressive regimes abuse such a state of emergency. South Africa in the 1970s and 1980s was a case in point. Instead, it would be more prudent to prevent the emergency from arising in the first place.[65] Thus, Clive Hamilton concludes, "If our failure to cut emissions is a result of the power of corporate interests, the fetish for economic growth and the comfortable conservatism of a consumer society, then resorting to climate engineering allows us to avoid facing up to social dysfunction, at least for as long as it works."[66] The attraction of technological fixes is precisely that they would solve a problem without requiring the messy complexities of social change. They therefore can at best address the symptoms of the underlying problem. In fact, it is because social change (in consumerist behavior) is regarded as *inconceivable* that recourse has to be taken to technological solutions.[67]

The hope that technology will save us is focused on *Homo Faber* and not so much *Homo sapiens*, to be clever but not necessarily to become wise. Hamilton comments, "Changing technologies always seem easier than changing people or challenging power."[68] With reference to Karl Jaspers he adds: "We want to find salvation in technological conquest—as if the human use of technology might itself be subject to technological directions."[69] For Hamilton, geoengineering may well be the most serious moral hazard humanity has ever faced.[70] This is pushed to the extreme by the hope that artificial intelligence may enhance human capabilities to calculate what is needed for human flourishing.

Toward the end of his book *We're Doomed. Now What?* Roy Scranton comes to the conclusion that Enlightenment thinking cannot save us from Anthropocene challenges. Despite all the intellectual efforts, in research and rhetoric, expressing grave concerns over climate change, we seem unable to act coherently and collectively to this threat, the greatest challenge the human species has ever faced.[71] That is the reason why "We're Doomed.", with a full stop, not a comma in the title. He observes that the Kantian urge to think for ourselves because it makes us free, that critical inquiry helps to make life better, that rational argumentation in the public sphere will promote better governance, that "think tanks" can resolve wicked problems, leaves the skeptical question unanswered: What is thinking good for—today, or ever before? In fact, he suggests that rationality offered Kant a way of expressing an abiding faith in a Christian God because reason (the Logos) comes from God, connects to God, and in some way is God.[72] With postmodern critics he recognizes the way in which thinking is embedded in power relations. Salvation in modernist capitalism was not derived from thinking but from colonialist exploitation and the imperialist plundering of resources. He then adds:

> Since thought has never been able to save us in the past, it might be a little unreasonable to expect it to do so now. Thought simply cannot solve the root problems of human existence—desire and mortality—and thought cannot tell us what is good, what we ought to do, or how to live, because these questions have no final answers. Thought cannot make us free, from nature or from each other, since human consciousness is determined by countless factors from gut bacteria, neurochemistry, and genetics to acculturation and the weather, and human politics is the product not of logical propositions but of situational negotiation within power relations. And finally, thought cannot make us gods. It cannot free us from animal, material existences, the demands of embodied life, and the fact of our finitude.[73]

This is not Scranton's final word though. He gives this another twist by saying: "Thought is, in the end, the only thing that *can* save us."[74] And: "Only thought can help us to see who we are, *know* who we are, and help us reconcile our imaginary collective models of the world, so rich and so vital and so often false, with the truth of our worldly being." This amounts to a Buddhist plea for humble yet diligent thinking that recognizes human dependence on other humans and nonhumans, that can interrupt the patterns of domination, and that can open the way to compassion and joyful communion with all other being.[75]

Yet it seems that failure is already built into the narrative of proposing geoengineering, given the argument that we have run out of time for alternative solutions.[76] If Plan A fails due to political inertia, we need to opt for Plan B in order to "buy time." If mitigation fails, then adaptation is required. If adaptation proves too difficult, then geoengineering is required. If land-use change and scaling down carbon emissions is unattractive, then CCS has to counter the continued use of fossil fuels. Indeed, the belief that carbon sequestration will save the world from global warming has become "something of an article of faith among policymakers."[77] However, if such technologies will at best become available only in decades, this will be too late to ensure a drastic reduction in carbon emissions. If all else fails, then solar radiation management is our last best hope, even though this is scarcely a permanent solution and has to be kept up (like an addiction) or else runaway climate change within a short time period will be the result (the so-called termination problem).

Is this not reminiscent of the failed ABC(D) strategy to combat the spread of the HI-virus? Abstinence (A) may be best, but if that does not work, be faithful (B) to your sexual partner. If that proves difficult for you, at least make sure to use condoms (C). If that is not done, face death (D) . . . or stick to antiretroviral treatment for the rest of your life. Moreover, the danger is that using condoms (solar radiation management) can provide a psychological rationale for sexual license (less commitment to carbon mitigation efforts). There is also the danger that arguments in terms of cost-effectiveness can become decisive: it is cheaper and easier to opt for sulfur spray (eating garlic to boost the immune system) than to mobilize around mitigation efforts (antiretroviral treatment), even though this does not resolve the underlying problem.

Although such technologies seem to suggest immense human power to manage even the atmosphere, the narrative framework within which this is embedded speaks of vulnerability and indeed human inability.[78] Indeed, the analogy with (male) sexual potency coupled with a feeble inability to control sexual desires is rather telling. Perhaps, as Laura Hartman also suggests, it may be better to opt for medical than engineering imagery to address the failure with mitigation efforts.[79] Accordingly, we need a proper "diagnosis" in order to find a suitable "remedy" or therapy for the "healing" of the atmosphere. But the patient will more readily regard the doctor's diagnosis as palatable if the disease is treated as curable, and if this does not require a change of habits or worldviews.[80] In other words, the available medicine of geoengineering techniques would be culturally congenial to help people accept the diagnosis of a changing climate. The facts have to conform to the values![81] Or is this where the medical and psychological imagery and its tendency to portray the "patient" as an innocent victim of circumstances beyond her or his locus of control also fall short? Perhaps the language of guilt, repentance,[82] forgiveness, and reconciliation, before God and the whole world, still goes deeper in order to avoid superficial solutions—in theological language "cheap grace."

Are there then no circumstances under which some or other form of geoengineering may be adopted? What if it is no longer a matter of preventing the further spread of the HI-virus, but treating the already infected, the ones with full-blown AIDS? Or, if the patient has cancer, is radical surgery and chemotherapy not advised even though this may have numerous side effects and may not be successful in any case? In other words, what if a climate catastrophe can no longer be avoided given the lag effect of carbon in the atmosphere? Should we not abandon the Promethean dream of finding a decisive solution and instead engage in the messy world of climate politics in order to avoid some of the worst symptoms?[83]

In a worst-case scenario: What if the disease is indeed terminal? What might comfort us, give us a sense of purpose, in the face of annihilation?[84] Would some form of treatment to ease the pain not be advised, even if frantic attempts to prolong the life of the patient are dropped? I will refrain from addressing such questions, also because the medical analogies would need further scrutiny. I would at least maintain that the worse the situation is, the less pretense of offering "salvation" may be endorsed. If death is unavoidable, even the death of a civilization, it is the art of dying (*ars moriendi*) that is required, not frantic forms of engineering or an overdose of morphine to ease the pain (religion as opium).

A FEW OTHER FORMS OF QUASI-SOTERIOLOGY

Arguably, one other false form of salvation is the system of international carbon trading through carbon offsets that was introduced as part of the original Kyoto Protocol. The system was designed to allow carbon emissions in national targets for mitigation as long as this is balanced by a reduction in carbon emissions in audited projects elsewhere in the world. Besides the many loopholes in the cap-and-trade system, abused by companies to suggest a reduction in greenhouse emissions, and fluctuations in the pricing of carbon, the system at best was designed to target increases rather than drastic cuts in carbon emissions. It is not surprising that the system is often likened to medieval practices to buy indulgences to reduce the length of one's stay in purgatory, or that of one's deceased relatives. Although, for pragmatists, this is at least a little better than doing nothing (and the carbon trading system could be reformed), it may well amount to inducing a guilty conscience for those amenable to it.[85] Otherwise, it may merely serve as a way for the consumer class to ease their conscience by paying for their carbon-intensive habits.

The hope to colonize other, distant planets in order to escape from this "doomed" planet scarcely deserves any discussion here, although it should certainly be included as a form of quasi-soteriology. This is nothing but a secular form of religious escapism. One is reminded of the parallels with the voyages of exploration and exploitation in "discovering" the "new world," claiming that the land was "empty" and available as resources for European nations—leading to the "Orbis spike" as one possible marker of the Anthropocene. This was repeated in the colonial "scramble" for African territories and resources leading to the "Berlin Act" of 1885, following the Industrial Revolution in Europe (another possible marker of the Anthropocene). The fallacy is not only related to the impossibility of such an escape, the elitist connotations (only the wealthy will go, the rest will be left to die), or the romanticism of such dreams. The problem is also that the rivalries between European states (those that are implicated in the Anthropocene) are merely exported elsewhere so that the underlying problems cannot be resolved through the occupation of new territories.[86]

There is yet another option for a quasi-soteriology that evokes "rumors of angels," that is, the hope that there may be other forms of intelligent life elsewhere in the galaxy that could offer us advice on how to live on planet Earth. This is a quite common strategy in science fiction: aliens are either warlike invaders or highly advanced as is evident in their mediating, peace-building initiatives. In a remarkable paragraph in *Earth in Human Hands*, David Grinspoon articulates such hope around the search for extraterrestrial intelligence:

> SETI is a hopeful enterprise. We are on one level of our massive double standard: that what we seek is something that has never existed on Earth, and is almost unimaginable different and more highly evolved than our own young, fragile, transitional civilization. In a sense, this search is an act of faith that such a historical outcome is possible. Scientists aren't supposed to believe in angels or miracles, but we can believe in the possibility of evolved ETs with godlike capabilities and unknown power to transform our species beyond our current troubled state.[87]

And on the hope for revelation:

> I maintain hope that the universe will show us a sign, that my (I tell myself) scientifically informed faith may be rewarded. . . . I've been waiting for a signal all my life, and although our powers of searching are just beginning to ramp up, I have to admit that I've felt my expectations drop somewhat.[88]

This hope seems to come full circle back to an interventionist view of Jewish-Christian hope. What, then, would an appropriate remedy be if the diagnosis is the intrusion of quasi-soteriologies? If climate engineering amounts to merely treating the symptoms in a palliative manner, it would be best to find a cure for the underlying disease.[89]

At the same time, the advent of the Anthropocene reminds us that the Earth's history has been characterized by many abrupt and dramatic changes. There are tipping points beyond which a return to what has been is no longer possible. The problem is not only that we do know where the tipping points lie but that we do not know what we do not know.[90] This poses hard philosophical problems regarding time—irreversibility, fatal delays, tipping points, inertia, and so forth.[91] The idea that nature is fundamentally stable, despite cyclical patterns, is derived from Platonic philosophy. Interpreted in Christian categories this assumes that God created the world to be very good, if perhaps not perfect. Although humans disturbed such stability, God's work of salvation is aimed at restoring that. It now seems clear that such a message is inadequate: the Earth's climate has been inherently unstable, subject to rapid change even in the absence of human activity.[92] Remarkably most forms of a quasi-soteriology seek to ensure that the Earth's climate remains in a relatively stable Holocene-like state. Those proposals that allow for and even encourage more dramatic shifts seem to be just too risky to even contemplate.

NOTES

1. See Durand, "A Confession – Was it Really Necessary?" 36.
2. The following three paragraphs draw from Coetzee and Conradie, "Apartheid as Quasi-soteriology," 120–22.
3. See Jay B. McDaniel, *Living from the Center: Spirituality in an Age of Consumerism* (St Louis: Chalice Press, 2000), 73.
4. See this comment by Peter Sloterdijk: "In truth, money has long since proved itself as an operatively successful alternative to God. This affects the overall context of things today more than a Creator of Heaven and earth ever could." See *In the World Interior of Capital*, 209.
5. See Goudzwaard, *Idols of Our Time*, 56. Goudzwaard describes the welfare state even as an idol, with rituals performed by a priesthood of politicians.
6. Serres, *The Incandescent*, 166.
7. Clive Hamilton, "Human Destiny in the Anthropocene," in *The Anthropocene and the Global Environmental Crisis*, edited by Clive Hamilton, Christophe Bonneuil and François Gemene (London and New York: Routledge, 2015), 32–43 (36).
8. See Morton, *Hyperobjects*, 21. Morton nevertheless places some hope in the advent of hyperobjects as good candidates for "the last god" (Heidegger), the saving power that grows alongside the dangerous power (Hölderlin).
9. Gonzalez, "Bridging the North-South Divide," 431.
10. See Haraway, "Staying with the Trouble," 46.
11. Wallace-Wells, *The Uninhabitable Earth*, 157.
12. *An Ecomodernist Manifesto*, 7.
13. See also Steffen et al., *Global Change and the Earth System*, 291.

14. *An Ecomodernist Manifesto*, 21.
15. See Hamilton, "Human Destiny in the Anthropocene," 39.
16. Hamilton, *Defiant Earth*, 48.
17. Ibid.
18. See, for example, my Ernst M. Conradie, *Saving the Earth? The Legacy of Reformed Views on "Re-creation"* (Berlin: LIT Verlag, 2013).
19. Hamilton, "Human Destiny in the Anthropocene," 40.
20. Hamilton, *Defiant Earth*, 155–56.
21. Ibid., 158, 159.
22. Clive Hamilton, "The Theodicy of the 'Good Anthropocene,'" *Environmental Humanities* 7 (2015): 233–38.
23. Hamilton, "The Theodicy of the 'Good Anthropocene,'" 236.
24. Bill McKibben, *The End of Nature* (New York et al.: Penguin, 1989); Carolyn Merchant, *The Death of Nature: Women, Ecology, and the Scientific Revolution* (San Francisco: HarperOne, 1980).
25. See, Serres, *Malfeasance*, 70. Serres is graphic: "Who doesn't see that the only thing left floating will be the homogenous excrement of the victorious Great Owner, Sapiens sapiens?" (70).
26. Hamilton, *Defiant Earth*, 48.
27. Ibid., 51.
28. See the Glossary as Annex II, edited by Katharine J. Mach, Serge Planton and Christoph von Stechow, to the Intergovernmental Panel on Climate Change's report entitled *Climate Change 2014: Synthesis Report*, 123, https://www.ipcc.ch/site/assets/uploads/2018/05/SYR_AR5_FINAL_full_wcover.pdf (accessed January 4, 2019).
29. I find the discussion by Clive Hamilton in *Earthmasters* sobering, to say the least. See also Forrest Clingerman and Kevin J. O'Brien (eds), *Theological and Ethical Responses to Climate Engineering* (Lanham: Lexington, 2016). Several of the essays in this volume employ terminology related to discourse on sin such as hubris (playing God, overreach), consumerist greed, and sloth (a failure to exercise responsibility, stewardship).
30. Sachs, *Common Wealth*, 100–2.
31. See also Jamieson, *Reason in a Dark Time*, 234.
32. Ellis, *Anthropocene*, 151.
33. The reference is to Paul J. Crutzen, "Albedo Enhancement by Stratospheric Sulfur Injections: A Contribution to Resolve a Policy Dilemma?" *Climatic Change* 77 (2007): 211–19.
34. Janos Pasztor, "The Need for Governance of Climate Geoengineering," *Ethics & International Affairs* 31:4 (2017): 419–30.
35. Ibid., 422. Pasztor's own conclusion is that a regulatory framework is needed, especially before any consideration can be given to solar radiation management: "As history indicates, nothing is totally predictable. But if current predictions about our inability to address rising temperatures quickly enough turn out to be accurate, the global debate over the availability and feasibility of climate geoengineering is about to switch into high gear. If it does, we need to be ready for it" (429). He nevertheless warns that "Whichever way we go, we will keep stressing that under no scenario should solar engineering be adopted as an alternative to what we already know how

to do. If it happens at all, it should be as a complementary approach to mitigation, adaptation, and greenhouse gas removal" (429).

36. On this I can agree with Mark Lynas, *The God Species*, 196. I would not support his use of pronouns when he adds, "*Our* hands are on the thermostat whether *we* like it or not, so sooner or later *we* are going to have to face up to the need to make a decision about what temperature we want *our* planet to be at over the longer term" (197, my emphasis).

37. The atheist public intellectual Clive Hamilton is one example of a scholar who questions whether it is wise "to try to play God with the climate." See Clive Hamilton, "Geoengineering: Our Last Hope, or a False Promise?" *New York Times*, May 26, 2013. See also his *Earthmasters*.

38. See the title of Mark Lynas, "Geo-engineering, Nuclear Power and Climate Change: Playing God is Good for the Planet," *The Telegraph*, July 12, 2011, an excerpt from *The God Species*.

39. See Forrest Clingerman and Kevin J. O'Brien, "Playing God: Why Religion Belongs in the Climate Engineering Debate," in *Theological and Ethical Responses to Climate Engineering*, edited by Forrest Clingerman and Kevin J. O'Brien (Lanham: Lexington, 2016), 3–15 (10). They also refer to the contrasting positions of Hamilton and Lynas.

40. Grinspoon, *Earth in Human Hands*, 168.

41. Ibid., 187.

42. Ibid., 188.

43. See Forrest Clingerman, "Time's Arrow and Narratives of Climate Engineering," in *Theological and Ethical Responses to Climate Engineering*, edited by Forrest Clingerman and Kevin J. O'Brien (Lanham: Lexington, 2016), 81–101.

44. See also Pinker, *Enlightenment Now*, 151.

45. Shellenberger and Nordhaus, *Love Your Monsters*, 2011, 9–10, as quoted in Bonneuil, "The Geological Turn," 26.

46. See also Pinker, *Enlightenment Now*, 153.

47. See Hamilton, *Earthmasters*, 116.

48. See Gardiner, *A Perfect Moral Storm*, 67.

49. For a discussion, see Gardiner, *A Perfect Moral Storm*, 339–96.

50. Steffen et al., *Global Change and the Earth System*, 289.

51. Crutzen, "Albedo Enhancement by Stratospheric Sulfur Injections."

52. See Steffen et al., "The Anthropocene: Conceptual and Historical Perspectives," 859.

53. Scranton, *Learning to Die in the Anthropocene*, 43.

54. Ibid., 51.

55. Hamilton, *Earthmasters*, 18.

56. Ibid., 155–56.

57. Ibid., 147.

58. Ibid., 167.

59. Hamilton, *Defiant Earth*, vii–viii.

60. Klein, *This Changes Everything*, 255. Elsewhere she comments, "the solution to global warming is not to fix the world, it is to fix ourselves" (255). She does not seem to notice her Pelagian assumptions here.

61. Haraway, *Staying with the Trouble*, 3. In the same breath she resists the assumption that it is "game over," that it is too late to even try to make anything better.

62. See Wood, *Deep Time, Dark Times*, 113.

63. Jonathan Symons, "Geoengineering Justice: Who Gets to Decide Whether to Hack the Climate," *Breakthrough Journal* 9 (2018).

64. Hamilton, *Earthmasters*, 171.

65. See Gardiner, *A Perfect Moral Storm*, 358.

66. Hamilton, "Geoengineering: Our Last Hope, or a False Promise?"

67. Hamilton, *Earthmasters*, 177.

68. Hamilton, *Requiem for a Species*, 180.

69. Ibid., 180. The reference is to Karl Jaspers, *The Future of Mankind* (Chicago: University of Chicago Press, 1961), viii.

70. Ibid., 186.

71. See Scranton, *We're Doomed. Now What?*, 311, 320.

72. See Ibid., 314.

73. Ibid., 315–16.

74. Ibid., 316 (his emphasis).

75. See Ibid., 317.

76. See Stefan Skrimshire, "Rewriting Morality: A Theological Critique of Geoengineering and De-Extinction," in *Theological and Ethical Responses to Climate Engineering*, edited by Forrest Clingerman and Kevin J. O'Brien (Lanham: Lexington, 2016), 103–26 (103).

77. See Hamilton, *Earthmasters*, 171, with reference to an article in *The Economist* of March 5, 2009, entitled "The illusion of clean coal."

78. See Clingerman, "Time's Arrow and Narratives of Climate Engineering."

79. See Laura M. Hartman, "Healing the Climate? Christian Ethics and Medical Models for Climate Engineering," in *Theological and Ethical Responses to Climate Engineering*, edited by Forrest Clingerman and Kevin J. O'Brien (Lanham: Lexington, 2016), 129–48.

80. Hamilton, *Earthmasters*, 90.

81. Ibid., 92.

82. See Kevin J. O'Brien, "'First be Reconciled': The Priority of Repentance in the Climate Engineering Debate," in *Theological and Ethical Responses to Climate Engineering*, edited by Forrest Clingerman and Kevin J. O'Brien (Lanham: Lexington, 2016), 187–207. O'Brien's argument is remarkable: we should neither accept geoengineering before repenting, nor should we reject geoengineering before repenting.

83. See Jamieson, *Reason in a Dark Time*, 238.

84. See Wallace-Wells, *The Uninhabitable Earth*, 175.

85. See Lynas, *The God Species*, 192. For Lynas, "Pragmatism beats purism. Every time."

86. See Latour, *Facing Gaia*, 232, in discussing the work of Carl Schmitt.

87. Grinspoon, *Earth in Human Hands*, 336.

88. Ibid., 338.
89. See Lynas, *The God Species*, 197.
90. Žižek, *Living in the End Times*, 429.
91. See Wood, *Deep Time, Dark Times*, 17–19.
92. Adapted from Angus, *Facing the Anthropocene*, 66–67.

Chapter 7

Idolatry in the Anthropocene?

APARTHEID AND IDOLATRY

Although apartheid was often described as a form of idolatry, this particular theological critique of apartheid was never really clarified in relation to apartheid as cultural habits,[1] structural violence, ideology, quasi-soteriology, or heresy.[2] It is almost as if any harsh word that could be used to denounce apartheid was added to the list. Does this imply that idolatry is hardly distinguishable from other critiques of apartheid? Would it make even less sense to approach an adequate diagnosis of what has gone wrong in the Anthropocene from this angle? Not necessarily.

If idolatry includes the connotations of admiration, adoration, reverence, worship, and glorification, the question would be what attracted such responses under apartheid other than the Triune God of Christianity. Note that it is the idolatry of Christians that are addressed here, not the idolatry of adherents to other religious traditions (akin to the Muslim notion of infidels). It would be inappropriate to name the heroes of apartheid since they were not self-consciously "worshipped," even if "hero worship" was found in the cases of Paul Kruger and Hendrik Verwoerd. Possibilities may include "whiteness," European civilization, Afrikaner identity (whatever that may mean), Afrikaans as language, or Afrikaner nationalism. While these were "admired," "adored," and certainly "glorified," reverence is perhaps too strong a word here, except in smaller fanatical circles. It may be closer to the truth to consider macho, high testosterone aspects of culture such as hunting (in an earlier dispensation), sport (rugby), and "braai" (barbeque), although the term idolatry is best used in jest here.

A critique of idolatry is perhaps epitomized by the document *The Road to Damascus: Kairos and Conversion* (1989) produced by Third World

Christians from seven countries, namely, South Africa, Namibia, South Korea, the Philippines, El Salvador, Nicaragua, and Guatemala. The document offers an extended critique of the legacy of imperialism, colonialism, and racism; analyzes the roots of the conflict between those who are economically powerful and the powerless; and issues a strong call for conversion to those Christians who side with money, power, privilege, and pleasure. Paragraphs 50 and 51 of the document focus on idolatry:

> In our countries, the worship of money, power, privilege and pleasure has certainly replaced the worship of God. This form of idolatry has been organised into a system in which consumerist materialism has been enthroned as a god. Idolatry makes things, especially money and property, more important than people. It is anti-people.
>
> Because the idol is anti-people, it demands absolute submission and blind obedience. The idols we read about in the Bible make their followers into slaves, prisoners or robots depriving them of freedom. Subservience to money dehumanises people. Profits are pursued at the expense of people. The graven image of the god of money is the security state that defends the system and demands absolute and blind obedience. In some countries, it is cruel and merciless; in others, it wears a deceptive mask. Those who disobey are punished brutally; those who obey are rewarded with material benefits and security. Idols rule by fear and intimidation or by trying to buy people, to bribe them and seduce them with money.[3]

IDOLATRY AND THE CRITIQUE OF CONSUMERISM

In religious critiques of consumerism, often although not necessarily in association with neoliberal capitalism, it is typically argued that consumerism amounts to a form of idolatry.[4] Such idolatry may be expressed in many forms, as worshipping the body, pleasure, excitement, money, and so forth. Indeed, idolaters are typically polytheists.[5] Of course, consumerism, hedonism, and materialism are seldom explicitly articulated as a form of religious belief. However, as Luther suggested in his large catechism, the object in which people put their faith is in fact their God. To rely on or trust an object other than God would therefore amount to idolatry, to finding or creating for oneself a surrogate god. This is also how the Heidelberg Catechism defines idolatry in its answer to Question 95: "Idolatry is having or inventing something in which to put our trust instead of, or in addition to, the only true God who has revealed himself in his Word." The problem is not an attachment to things or caring for them excessively but putting one's trust in something that is not God. Another way in which idolatry can be detected is in terms of that for which people are willing to make sacrifices.[6] Mammon typically demands human sacrifices: "The only one that we are willing to sacrifice everything for, is indeed our God."[7] Or, to put this in other terms, "We no longer give our surplus to God; the process of producing an ever-expanding surplus is in itself our God."[8]

Consumerism may therefore not only be described as an ideology, but indeed as a form of (civil) religion. Writing from a Buddhist perspective David Loy claims that the Market is "already the most successful religion of all time, winning more converts more quickly than any previous belief system or value-system in human history."[9] He adds that market capitalism "is becoming the first truly world religion, binding all corners of the globe more and more tightly into a worldview and set of values whose religious role we overlook only because we insist on seeing them as secular."[10]

From a Christian perspective William Stringfellow adds that where money becomes a form of idolatry, to be poor is considered to be morally inferior, while money is considered as evidence of the moral worth of a person. Moreover, while the poor die and are soon forgotten, the estate of the rich can purchase for them a form of immortality. According to this form of idolatry, money therefore makes people not only moral but also immortal.[11]

The biblical notion of "Mammon" typically provides a point of entry for such a critique of consumerism. The contention here would be that the desire for what money can buy has functionally replaced faith in God. Something that is of some worth is being treated as if it were of ultimate worth. We live as voluntary slaves under the reign of the infrastructures of Mammon.[12] This is the argument of the Dutch reformed economist Bob Goudzwaard. He maintains that the quest for economic growth and material progress has become idolatrous, in that people in industrial societies have become so obsessed with becoming more affluent that they are motivated more by this than by any relationship with God. When the finiteness of economic categories such as human needs, the market, and economic growth is denied, such categories become ideological and even idolatrous.

Goudzwaard analyzes the religious nature of the global economy in terms of three examples, namely: (1) the way in which insatiable human needs are taken for granted; (2) the assumption that economic distortions can be rectified through the functioning of the all-pervasive Market; and (3) the way in which economic growth is portrayed in salvific terms as a panacea for poverty, social ills, and human happiness, albeit that it also demands certain sacrifices to ensure such growth.[13] He refers to the booklet by Andrew Carnegie significantly entitled *The Gospel of Wealth* in which he praises our "wonderful material development, which brings improved conditions in its train" and which formed the basis of his own philanthropy. Carnegie adds: "Such, in my opinion, is the true Gospel concerning Wealth, obedience to which is destined some day to solve the problem of the Rich and the Poor, and to bring 'Peace on earth among men of Good Will.'"[14] Goudzwaard notes the religious language employed and adds that such idols not only distort life but eventually enslaved those who put their trust in them. We are enslaved by our belief in economic growth and progress. And, as Clive Hamilton adds, this may lead to radical distortions—to the point that the governments of the world become willing to sacrifice the planet to the supreme god of economic growth.[15]

American process theologian John Cobb argues along similar lines that what he calls "economism" "has become the global religion of our time, that it is idolatrous, and that Christians should name its god, wealth, as an idol."[16] He describes "economism" as a system according to which the economic values of neoliberalism are viewed as primary and therefore as determinative of domestic, national, and international policies. He identifies the following implicit claims made on this basis: (1) that an increase of wealth can overcome poverty without class struggle, (2) that sustained economic growth is necessary to address unemployment, (3) that economic freedom tends to translate into the exercise of democratic freedom, (4) that economic growth is the key to address population growth, (5) that economic growth is necessary to address environmental degradation, and (6) that economic growth helps people to pursue cultural and other values. Together these claims constitute a powerful argument for "worshipping" Mammon.[17] Cobb acknowledges the elements of truth in these claims but argues that these are indeed half-truths that hide the ways in which moral communities and community leadership are being undermined by globalization. Construing half-truths is typical of idolatry.

This is also the argument of the German theologian Falk Wagner in his book *Geld oder Gott* (1984). Wagner uses the term "money pantheism" to describe the way in which Mammon turns everything else into means toward the ultimate end of obtaining more and more money, power, status, and pleasure. Capital becomes not only one form of interpersonal communication through trade, but an absolute form of communication by which all other forms of communication are governed, including culture, art, and religion (following insights from Niklas Luhmann). Mammon thus becomes an end in itself, an omnipresent and almighty God who is the origin of everything that is valuable. Following Luther, Wagner suggests that people tend to put their trust in financial insurance and security, and no longer in the Way of Jesus Christ.[18] In this way, capital becomes a functional equivalent of God in ordinary life. In response, Wagner offers a critique of any form of self-grounding religion and a retrieval of the Christian faith in the Triune God on the basis of God's self-disclosure in Jesus Christ. Wagner thus offers a theological and not merely an ethical critique of a capitalist economy (and of consumerism).[19]

Drawing on Wagner's analysis, the Church of Norway in its remarkable report on consumerism observes that "the power of money could work as God, because it too can become an almighty, all-embracing hidden reality which directs and conditions all human relations. The power of money to promote and secure consumption, becomes, along with the market, the invisible power which rules everything, and which is the basis for all human ways of communicating, and for the organization of human communities."[20]

Michael Northcott argues that in a consumer society, at a fundamental level, salvation is expected to come through the functioning of the Market which may be contrasted with salvation by the "central plan" in communist discourse. He comments:

> Global warming is the earth's judgment on the global market empire, and on the heedless consumption it fosters. The neoliberal claim is that the market, combined with technological power, can redeem the peoples of the world from pain and suffering through the autonomous, self-regulating market system. Those who direct the neoliberal project of economic globalisation presume that the welfare of people and planet is advanced when the most powerful economic actors—multinational corporations, bankers, investors, engineers—are freed from regulation and taxation to make and sell more "consumer objects" and accumulate more wealth.[21]

He adds:

> Neoliberalism rests on a collective lie of momentous proportions, which is that when economic growth is pursued without let or hindrance, and when material power is concentrated in the hands of economic corporations, the peoples of the earth will enjoy progress and peace.[22]

Even if belief in not only a transcendent but also an immanent God declined through secularization, a series of successors emerged to replace God, including the nation-state and the Market.[23] Harvey Cox comes to the conclusion that the Market has assumed the same character traits as God, namely, omnipotence (the capacity to define what is real, the power to create something out of nothing), omniscience (the market is able to determine what human needs are, what the price of commodities and services should be, what salaries are appropriate), omnipresence (touching on all aspects of life, everything can be commodified, including spirituality), and benevolence (it will ensure well-being for all).[24] He adds, "Likewise, although The Market, we are assured, possesses these divine attributes, they are not always completely evident to mortals but must be trusted and affirmed by faith."[25] Either way, the market is easily personified as "nervous," "bullish," even "jubilant."[26] Cox adds that the god of "The Market" is no longer merely one among many in a pantheon of gods: "The Market is becoming more like the Yahweh of the Old Testament—not just one superior deity contending with others but the Supreme Deity, the only true God, whose reign must now be universally accepted and who allows for no pretenders."[27] Cox evidently enjoys the exercise of exploring the many analogies between religion and the market—complete with priests, rituals, saints, heroes, prophets, codes, and heresies. He has no doubt that The Market operates as a surrogate religion and that this is evident from the way criticism against The Market is deflected and discouraged.[28]

Such a critique of the Market is also found in ecumenical literature. The South African Council of Churches' statement on climate change has this:

> Several surrogates for trust in God have emerged: political power, quality education, access to swift technology, the power of positive thinking, marketing and, perhaps above all, the Market. The Market has assumed several divine characteristics. It is supposed to be benevolent, invisible and omnipresent, even omnipotent in order to produce wealth and distribute benefits, costs and obligations to all. We are called to trust that the Market will indeed deliver, even where these results are not yet obvious. This even poses a consumerist version of the theodicy problem: Why does an omnipotent, omnipresent and omniscient (if invisible) Market fail to deliver well-being to everyone, if it is in principle benevolent? Why do bad things happen to poor people if the Market should ensure that resources are distributed more or less equitably?[29]

In his Apostolic Exhortation *Evangelii Gaudium* (2013), Pope Francis responds to economic inequality and financial meltdown by saying not only "no to an economy of exclusion" (as quoted above) but also "no to the new idolatry of money." Two paragraphs may be noted here:

> 55. One cause of this situation is found in our relationship with money, since we calmly accept its dominion over ourselves and our societies. The current financial crisis can make us overlook the fact that it originated in a profound human crisis: the denial of the primacy of the human person! *We have created new idols. The worship of the ancient golden calf (cf. Ex 32:1-35) has returned in a new and ruthless guise in the idolatry of money* and the dictatorship of an impersonal economy lacking a truly human purpose. The worldwide crisis affecting finance and the economy lays bare their imbalances and, above all, their lack of real concern for human beings; man is reduced to one of his needs alone: consumption.
>
> 56. While the earnings of a minority are growing exponentially, so too is the gap separating the majority from the prosperity enjoyed by those happy few.

> This imbalance is the result of ideologies which defend the *absolute autonomy of the marketplace and financial speculation.* Consequently, they reject the right of states, charged with vigilance for the common good, to exercise any form of control. A new tyranny is thus born, invisible and often virtual, which unilaterally and relentlessly imposes its own laws and rules. Debt and the accumulation of interest also make it difficult for countries to realize the potential of their own economies and keep citizens from enjoying their real purchasing power. To all this we can add widespread corruption and self-serving tax evasion, which have taken on worldwide dimensions. *The thirst for power and possessions knows no limits. In this system, which tends to devour everything which stands in the way of increased profits*, whatever is fragile, like the environment, is defenseless before the interests of a deified market, which become the only rule.[30]

One may conclude that idolatry thus suggests not only putting trust tacitly in false saviors but explicitly praising (even worshipping) something as divine (or quasi-divine) that is the product of our own imagination and construction. This is confirmed by the Greek etymology of idolatry coming from *eidolon* (image) and *latreia* (worship). From a theological perspective, one may say that idolatry is "the misdirection of the love, faith and devotion that are properly God's to that which is not God."[31] It is to treat as ultimate something that is penultimate, or even mundane. To expose such idolatry it is necessary to name such idols. What forms of idolatry may be found in the Anthropocene?

SYMPTOMS OF IDOLATRY IN THE CONSUMER SOCIETY

There is an obvious link between the emphasis on the "great acceleration" in discourse on the Anthropocene and the consumer society. To identify forms of idolatry in the Anthropocene it may therefore be helpful to sketch some of the "symptoms" of the underlying "disease" that manifest itself in the consumer society, drawing on my earlier work on *Christianity and the Critique of Consumerism*. There are indeed many symptoms of idolatry in a consumer society:

Evy McDonald comments: "Our affluence has shifted our focus from consuming only what we need to consuming for the sake of consuming. Consumption has become a habit, a hobby, a sport."[32] Indeed, an often-quoted survey of teenage girls in the United States discovered that 93 percent of them identified store-hopping as their favorite activity.[33] The one place where one can be assured of a welcome is in a store. Consumption has thus become a crucial determinant of cultural identity: "I shop therefore I am." This creates a

Faustian world where there are no limitations placed on desiring, where innumerable desires can all be satisfied—on condition that one surrender one's own soul.[34] Hans Küng adds that consumerism constitutes a religion of prosperity and progress, with the "gods" of sex, car, and career operating as a holy Trinity for conformists.[35] Ted Peters observes that the insatiable demands of the affluent leads to the habit of moving sponge-like through life, sopping up one experience after another, soaking in rich food, marvelous media spectacles, exciting travel to ever more remote destinations, thrilling risks, giant explosions,[36] the ecstasy of speed,[37] gambling on the stock exchange, sexual pleasure, and even religious ecstasy.[38] Paradise on Earth is found where the problem of scarcity is overcome through an excess of consumer goods, where all desires are satisfied and where no sacrifices are required.[39]

However, the attempt to accumulate such experiences correlates with an inability to assimilate such experiences since everything is done in a hurry, in order to "keep up."[40] Moreover, a repetition of an experience of pleasure soon fades away to a feeling of mild contentment and then to boredom. As a result, hedonism can become a form of opiate to relieve boredom. Sallie McFague comments: "We enjoy the consumer lifestyle; in fact we are addicted to it, and, like addicts, we cheerfully stay in denial if possible."[41] This is also Mary Grey's assessment:

> But in psychic and psycho-spiritual terms, what is destructive is that this addiction has hi-jacked our imaginations and cheapened and vulgarized our aspirations of fulfilment, the degree of mutuality and intimacy we could attain. It has substituted an insatiable, endless grasping for some new consumer good for yearning for the infinite and experience of the sacred.[42]

Peter Sloterdijk, in a remarkable formulation diagnoses such boredom in terms of an eroding of any notion of good and evil. Evil becomes a mere mood:

> In the climate of universal comfort, what was once original sin comes to light as trivial freedoms to do evil. In addition, once evil is deprived of its historical pretexts and utilitarian facades, its can only crystallize into its quintessential form in the state of post-historical boredom (*skuka*): purged of all excuses, it now becomes apparent—perhaps surprisingly for the naïve—that evil possesses the quality of a mere mood.[43]

The consumer culture with its depletion of available resources, its production of waste, and its quest for pleasure and happiness cannot hide an inner spiritual emptiness. As Douglas John Hall observes, the rhetoric of progress, mastery, and success which has elicited a "programmed optimism of the affluent consumer-driven West" is being undermined by a sense of fatalism over instabilities such as terrorism, prospects for economic collapse, deadly diseases, and atmospheric change.[44] Since consumerist extravagance cannot ultimately provide happiness or security, there emerges a new curiosity toward what Segundo Galilea calls the "superstitions of hope": horoscopes, telling the future, charms that should protect the house, a false confidence in technology and the sciences, in the power of psychology and the medical sciences to provide healing, in the substitutes of political ideologies, and so on.[45]

Such a sense of spiritual emptiness may also prompt a quest for spirituality in an age of consumerism.[46] Ultimately, consumerism cannot satisfy the deeper yearnings of human spiritualities—for that which is larger than ourselves, that which is infinite, that which is sacred. However, there is also a danger that spirituality itself may be commodified. There is indeed a burgeoning market for publications on spirituality. In fact, the quest for warm spirituality itself may be viewed as symptomatic of the emphasis on experience in consumerist cultures. There is a hunger for religious experience but an aversion to any theological definition of such experience. Gregory Jones argues that such a yearning for spirituality "is shaped by consumer impulses and captive to a therapeutic culture."[47] He comments: "The bifurcation of spirituality and politics is one of the reasons that contemporary spirituality can be consumed as a luxury consumer good, primarily designed for middle- and upper-class folks."[48] It is all too often cutoff from theological reflection and from sociopolitical realities. This may hinder rather than assist attempts to retrieve and embody authentic Christian practices. Jones argues that such literature "is dangerous because its invocation of 'spirituality' suggests that it will help readers nurture and discover authentic relationships with God, when what it offers is a synthetic substitute of vague, self-referential religiosity."[49] For Jones, the problem is that, in the quest for self-knowledge and self-expression, the self becomes the source, referent, and criterion to judge spiritual authenticity. By contrast, in the works of many Christian mystics the clue to knowledge of the self is being addressed by and known by God.[50] This is also reflected in the enigmatic opening lines of Calvin's *Institutes* with its acute hermeneutical sensitivity: "Nearly all the wisdom we possess, that is to say, true and sound wisdom, consists of two parts: the knowledge of God and of ourselves. But, while joined by many bonds, which one precedes and brings forth the other is not easy to discern."[51] In modernity, the assumption may have been that knowledge of the self can be obtained without reference to knowledge of God. In the Anthropocene, some may assume that knowledge of the self is itself knowledge of God.

IDOLATRY IN THE ANTHROPOCENE?

Such a critique of idolatry in the consumer society, in terms of the production, distribution, and consumption of wealth, is clearly still relevant in the Anthropocene. This is also found in secular discourse. An early precursor may be found in the words of Rachel Carson: "The modern world worships the gods of speed and quantity and of quick and easy profit, and out of this idolatry monstrous evils have arisen."[52] One may also comment on the gods of Nature, Evolution, Reason, Market, State, Development, Communication, and Media attention that typically assume all the godlike characteristics and creative capacities that God the Creator is denied by secular critics. Likewise, growth becomes both the supreme end and the supreme means for achieving salvation.[53] Indeed, as Susan Neiman recognizes, "the cunning of reason, the invisible hand, and the proletariat are different ways to replace the hand of Providence and cannot be understood without it."[54] Yet, it is far from clear whether idolatry is an appropriate conceptual tool to explore what is wrong in the Anthropocene, also from a secular perspective.

Arguably, the critique of idolatry has become a relic of the past in pluralistic cosmopolitan contexts where instead of the critique of worshipping false gods, it is the failure to tolerate, to respect, even to appreciate religious diversity that is subject to critique. There is indeed a reversal of values, symbolized by the title of Friedrich Nietzsche's *Twilight of the Idols* (1888), so that "Idols" may be promoted and turned into celebrity television programs. Social cohesion in a religiously plural society can only be found on the basis of religious tolerance. Any critique of idolatry, especially if aimed at outsiders, smacks of the kind of intolerance that will return us to the religious wars of the past. Monotheism is often regarded as dangerous, prone to fundamentalist abuse, while religious pluralism (or relativism?) may be regarded as the default position. Mark Sagoff, an apologist for ecomodernism, argues, for example, that ecomodernists should reject the monotheistic view of nature that environmentalists borrow from Judeo-Christianity, namely, that when God became One, nature became one. He even suggests that environmentalism has been described as "Calvinism without God." By contrast, pre-Christian religions viewed nature as local and assigned gods and spirits to trees, the wind, the harvest, and the sun. Sagoff concludes, "The theological hope of ecomodernism is that we can understand nature to be many, many places, each with its own guardian spirit. The hope is that human beings will become the guardian spirits of the natural world."[55]

Paradoxically, religious pluralism may well become "the only way"—in order to resist any forms of exclusivism or even inclusivism that may wish to include the "other" within the self. If such inclusivism is not a form of colonizing it is at least an *Anschluss*. In a so-called secular age, forms of agnosticism and atheism are widely prevalent although militant (missionary) atheism may be critiqued for the same exclusivist intolerance found in monotheism. For many citizens in industrialized societies, polytheism, recovering forms of animism, and exploring the female deities of primal religions seem far more attractive (and esoteric) than the muscular religions of the Axial age and their organized forms in world religions. As Susan Neiman observes, even the sins ascribed to pagan gods are worthier of imitation than that of monotheism: "What are lust and adultery next to the cruelty and vengeance ascribed to the Christian god of love."[56] The default assumption is therefore a rough equality or commensurability between most forms of religion,[57] while only fundamentalism, religious support for terrorism, and extremist forms of Satanism may still be denounced as intolerant. The assumption of religious freedom is that religion is a matter of voluntary association and private worship, so that the public influence of religion is necessarily constrained.[58] In short, if "God" is no longer needed as a working hypothesis for the production of wealth or social stability, the critique of idolatry seems misplaced.[59] The Dutch ethicist Hans Dirk van Hoogstraten formulates this provocatively: "In terms of the Holy Trinity, we could say that the place of God the Father is taken by the economy, the place of the Son by politics, and the place of the Holy Spirit by ethics. The Son (politics, government) executes the will of the Father (economics, free market), and they both send the Spirit (ethics, ideology) to the people. The Spirit provides social directives and ethical rules."[60] Either way, the "voice of God" represents the moral order of the Western elite, more recently of the consumer class.[61]

To reintroduce the critique of idolatry in such a context would not only seem out of place but would also symbolize a new spirit of religious intolerance. It would be as if the experiment of religious tolerance once again failed—as it did in 1610 when Henry IV of Navarre was assassinated—leading to thirty years of war from 1618 to 1648.[62] Such a dispensation is not inconceivable in the Anthropocene, given the return of identity politics, militant, and fascist movements, suggestions that a form of dictatorship may be the only way forward to address grave climate dangers amid the collapse of markets, and the popularity of apocalyptic imagery. In other words, if the Anthropocene would welcome back a critique of idolatry, this would not help with an adequate diagnosis but would itself be indicative of the underlying problem.

Does this imply that the critique of idolatry is a dead end? Perhaps . . . . However, such a critique requires a clear distinction between the modern quest for objective truth in science (even if in the form of the theory of relativity) and the willingness to live with plurality and ambiguity in culture and religion. Indeed, "you shall under no circumstances make knowledge of the laws of nature commensurable with any cult!"[63] Such a clear distinction between nature and religion is one that has become troubled in the Anthropocene. Yet, there remains an appeal to supreme authority to judge truth from falsehood both in science and in the critique of idolatry. Bruno Latour captures the irony here:

> Even those who violently reject the monotheistic religions have borrowed from them this quite particular way of violently rejecting idolatry. Iconoclasm is our common good. From the true God fulminating against all idols, we have moved to the true Nature fulminating against all the false gods. The division has remained, as have the lightning, the thunder, and the smell of the storm.[64]

What makes this awkward is that substitutes for God as a super-agent is replaced by categories such as Nature, Evolution, Society, the Market, or the System that are devoid of such agency.[65] Thus, things that are less than human are exalted as divine—a strategy that is typically exposed in the prophetic critique of idolatry.

Moreover, in naming the Anthropocene as such a new candidate for such idolatry has emerged, namely, the "God species" (Mark Lynas), also called *Homo Deus* (Yuval Harari). Through the rise of modern science we are even able to exercise God's power, to bring forth life based on chemical algorithms.[66] Indeed, humans are now like gods: they not only desire favorable weather (for social comfort, good harvests, and to maintain economic and military interests), but also hope to have the technical means to bring that about.[67] This is the hope of humans to achieve immortality, to become divine. Accordingly, we might as well abandon the lesser gods of money, status, and consumption, and indeed the celestial God. Humans, or rather some humans, perhaps trans-humans, may hope to become not only the "masters of the planet" but through artificial intelligence, genetic engineering, the dream of extended longevity, geoengineering, and space travel (even time travel) also the masters of the universe. In the words of Harari, "having raised humanity above the beastly level of survival struggles, we will now aim to upgrade humans into gods, and turn *Homo sapiens* into *Homo deus*."[68] Harari observes that it may be impossible to maintain social order without meaning but, against all expectations, God's death did not lead to social collapse. Humans were able to find meaning in themselves without reference to some great cosmic plan, to create meaning in a meaningless world.[69]

In *The God Species* Mark Lynas explores the myth of Prometheus who stole fire, the preserve of the gods, from the supreme god Zeus and brought it to humans. He comments that the human use of fire has given us an evolutionary advantage through cooking food (releasing more energy for human brains to develop), protection against predators, and warmth during cold nights. This food-fuel relationship defines the fire-ape, *homo pyrophilus*. At the advent of the Anthropocene such use of fire became expanded through the use of fossil fuels for transport, industry, and technology. Lynas comments that "being armed with fire put the rest of the world at our mercy" and adds, "Using the tool of the gods, we were to become as gods."[70] For Lynas, the God species should not shirk from such responsibility but should learn to use such divine power wisely.

In this way the decentralization of the place of humans in the cosmic scheme of things (see above) has become inverted. Following the shift from a geocentric to a heliocentric worldview (Galileo), the recognition of human descent in common evolutionary history (Darwin), the acceptance of a geological timescale (deep time), and the exploration of exoplanets and perhaps astrobiology, humans have become decentered in cosmic and evolutionary history. Nevertheless, human societies are the most complex feature of the universe that we know of. With the recognition of the Anthropocene it seems that we are back in the center of the story—prompting considerable confusion.[71] The prophet of such Promethean aspirations was Friedrich Nietzsche. God had to be declared dead so that the *Übermensch* can come to fruition—now in the shape of the cosmopolitan consumer.[72] The philosopher Susan Neiman traces this back to the failure of the theodicies provided by philosophers such as Leibniz but also Rousseau and Kant. With a parody on Marx's famous thesis, she comments: "Theodicies had hitherto tried to defend God, the point was to replace Him."[73] She explains:

> The wish to displace God that is contained in every attempt to re-create the world is the very essence of the sin of pride. It's pride that can lead to rebellion caused by the contemplation of all the evil in Creation. If God failed to get it right, why don't we do without Him and take over the job ourselves. The urge to humility is a product of acquiescence, if not terror: we agree not to understand why there is evil.[74]

In the context of the Promethean dreams to control the Earth's climate, the Enlightenment critique of religion, namely, that humans create their gods in their own image, according to their needs, desires, and aspirations, becomes radicalized. As Ludwig Feuerbach already asserted, "The turning of history will be the moment when man becomes aware that the only God of man is man himself. Homo homini Deus!"[75] We become gods when we are able to create "God" in our own image. This is no longer meant as a critique of religion but as a celebration of human ability. We can not only "make" God but custom engineer God to ensure that there is a space for a system-compatible God.[76] This becomes a matter not only of "playing God" but of becoming divine, with godlike attributes. Athanasian orthodoxy is radicalized: God in Jesus Christ became human so that we can become divine (instead). To gain human autonomy it is not enough to imitate or abandon God; God must be killed so that we can become divine.

Erle Ellis cheerily declares that the Anthropocene is far from the crisis it is portrayed as; instead it offers a new beginning, "ripe with human-directed opportunity."[77] He explains that "our unprecedented and growing powers also allow us the opportunity to create a planet that is better for both its human and nonhuman inhabitants. It is an opportunity that we should embrace."[78] Elsewhere he adds,

> We are poised at an important time in human and Earth history. For the first time, we have clear knowledge that we can and are changing the way the entire planet functions. This is an amazing opportunity—humanity has now made the leap to an entirely new level of planetary importance.[79]

Ellis likens this to parental responsibility and asks, "Will we be proud of the planet we create in the Anthropocene?" He answers that he considers it his duty to try to make this happen. He admits that "the Earth system we are creating won't be any more perfect than we are," but bets that the good will outweigh the bad.[80] The use of the word "create" is astonishing. It seems that humans are no longer merely cocreators with God, but are able to create the planet themselves, assuming the task of an "almighty Father," the Creator of heaven and Earth. In the words of Stewart Brand often quoted by Ellis, "We are as gods and HAVE to get good at it."[81] To which Mark Lynas adds an "Amen to that"! There is a similar tone of optimism in David Grinspoon's assessment that we as human beings are able to "play god" as far as mass extinctions are concerned. He says:

> For our entire history, human beings have been causing brutal extinctions of other species, but here's a hopeful thought: twenty-first-century humans could be first ever to decide *not* to behave like this. From now on, or as long as humans (or thinking creatures descended from or created by us) are here on the planet, extinction will be a choice. Obviously the current carnage, the mass extinction that we are now starting to manifest, has to stop. Then what? What extinction rate would we prefer? Do we wish to eliminate it entirely? Then what about biological innovation? If we choose to eliminate all extinction, then this is equivalent to saying we are the gods in charge of all future species. We will either have to play god or allow species to go extinct (which, I suppose, may still be a form of playing god).[82]

If and when such a God will prove to have feet of clay, this may well invite apocalyptic imagery. Then *homo industrialis* will be turned into a fallen angel, even the Antichrist who is bringing on the "end of history."[83] Or, in the words of Donna Haraway,

> The Story of Species Man as the agent of the Anthropocene is an almost laughable rerun of the great phallic humanization and modernizing Adventure, where man, made in the image of a vanished god, takes on super powers in his secular sacred ascent, only to end in tragic detumescence, once again.[84]

Such a narrative of the fallen hero poses a remarkable problem for a diagnosis of what has gone wrong in the Anthropocene. Does the problem lie with human nature or with human error? Mainstream economics assumes that human beings desire well-being, that self-denial is counterintuitive and won't work, and that growth in consumption is therefore aligned to human nature and becomes nonnegotiable.[85] If so, then there is something in human nature that becomes self-destructive under the conditions of the Anthropocene, that is, with a large human population, with high levels of consumption and sophisticated technology (I=PAT). It is hard to see where the remedy would then lie, short of changing human nature. Through genetic engineering? However, such a Promethean dream of altering human nature may well itself be nothing but a symptom of the underlying problem.

If, on the other hand, the problem lies with human error and if such error has taken the shape of a quasi-religious belief system in the form of a fetish, an addiction, a psychological obsession, a neurosis, even a spell, then it seems that what is needed is nothing short of a religious conversion.[86] For the ecosocialist critic Ian Angus this means that political leaders and economic leaders have to recognize that the gods they now serve are false.[87] However, in secular discourse, especially in a Marxist context, this is hard to come to terms with. Is a different ideology needed? Should material conditions be altered? Will the Earth itself ensure that a new equilibrium be reached? Where would help come from?[88] Timothy Morton, for one, is driven to agree with Heidegger that "only a god can save us now," but he comments that Heidegger forgot to add, "We just don't know what sort of god."[89] These are indeed the final words of his book *Hyperobjects*.

Either way, if the problem is indeed human self-idolatry, it seems clear that the problem needs to be addressed at that level too. In *Requiem for a Species*, Clive Hamilton recognizes that the modernist pursuit of utopia through endless growth, based on the belief in the human ingenuity, in the power of knowledge, and in the human capacity to control things, cannot be sustained.[90] The lesser gods of money, growth, and hedonism have to be abandoned. The powers we most trust to save us (freedom, democracy, scientific progress, technology) have betrayed us and now threaten to devour us.[91] Somewhat surprisingly, he looks toward the celestial gods to retaliate, to reinforce a sense of cosmic mystery. Perhaps, he concedes, it is only the celestial creator God who has the power to save us.[92] Perhaps we will need to turn to a sense of the sacred for protection.[93] He even asks, "Will the late surge of militant atheism come to be seen as a Homeric burst of pride before the fall?"[94]

In *Earthmasters*, Hamilton devotes a section to the dangers of "playing God" and tests the attempt to mimic God's omniscience, omnipotence, and benevolence through geoengineering. Even atheists aspire to attain such Godlike attributes. Hamilton critiques such a pursuit with the astute observation that we want "to supplant the gods in order to counter the mess we have made as faulty humans."[95] The danger is not only the dire consequences that may follow from the Promethean attempt to enter the domain of the gods, that playing God "betrays a deep fault in the human character."[96] He regards it as safer to face up to our failures and to devote ourselves to the Pelagian task of becoming better humans. Remarkably, he recognizes the ecological significance of maintaining a distinction between Creator and creature:[97]

> According to this Soterian view, if we are so mistaken in our understanding of the world and our role in it that we are drawn into playing God with the future of the planet, then thinking must be grounded in a different relationship between humans and the world, one that recognizes the boundary between the domain of mortals and that of the gods.[98]

Such Pelagian attempts at self-improvement seem to pose less of a problem in Augustinian theology given the classic distinction between human nature and fallen human nature. Humans were created by God and are deemed to be "very good," but have fallen from grace. Admittedly, the plausibility of the "myth" of such a "fall" is highly contested, also within contemporary Christian theology, to the point that some have asked, "Should we drop the fall?"[99]

This requires not only a retrieval of Christian symbols but also a revisiting of the "myths" embedded in the Jewish–Christian tradition. My sense is that Jung Mo Sung is on the right track here.[100] Following Franz Hinkelammert he refers to the belief, deeply embedded in Orthodox forms of Christianity, that because God in the incarnation of Christ became human we humans can become divine. For Hinkelammert, this process of divinization is indeed the foundational myth of the modern West: humans must become gods.[101] If the attempt to overcome human limits was previously regarded as hubris, in modernity it is regarded as a virtue since there can be no limits to human development. Moreover, although modernity has rested upon the denial of divine transcendence, this has been expressed in terms of its inverse, namely, the denial of human limitations, while the quest for the divinization of humans and the building of heaven on Earth remains intact. The god of the Market cannot accept its ecological limits either. It must expand endlessly.[102] For Sung the underlying problem remains treating what is finite as infinite. He insists that God is not to be found in material objects or in temples and concludes, "We human beings cannot attain the infinite (to see God) but we are able to experience the presence of God among us, to find fulfilment of our infinite desires, as far as is possible, in relationships of love and solidarity."[103]

Given such a critique of idolatry and a functional replacement of faith in God by substitutes, it seems to me that the question is not whether one believes in God or not, but in which God/gods one believes. In a pluralistic culture where it seems a matter of consumer choice to select a God that can best meet one's spiritual needs, many would wonder why they need to pick and choose. Would some form of eclecticism not make better sense? Is it not strategically wise to alternate one's gods depending upon the circumstances? Why back only one horse?

For Christians this is hardly an option. As Joerg Rieger argues in his book *Jesus vs. Caesar* there is a fundamental tension at the heart of Christianity.[104] This is not a tension between belief in God or atheism or secularism. Nor is it a choice between organized religion and spirituality, or even between Christianity and other religions. The choice is between seeing Jesus Christ or Caesar as the manifestation of God. The question is not whether or not it is absurd to claim that Jesus has *divine* status, but whether it is absurd to claim that this *Jesus* could be divine. The question is not whether Jesus is like God but whether God is like Jesus.[105] The contrast is indeed stark: Caesar was the embodiment of imperial power, while Jesus died as a poor peasant, a construction worker, in a remote province, without children, possessions, writings, or even followers (except for his mother and a prostitute). For Rieger, this is a tension between a faith that is inclusive and life-giving for all, especially those who are marginalized by contemporary constellations of power, and a faith that is destructive for people and the Earth. In short, it is a matter of choosing the right God, a message that is indeed good news for people who are tired of serving the wrong God, as Rieger's subtitle claims. Imperial Christianity opted not to make such a choice and to reconcile the Christian message with imperial power. For Rieger, it matters how and where one sees God as active—or else one may end up seeing God as an elitist dictator, a heavenly bully, or a supernatural control phreak.[106]

There may be one thing that is worse than idolatry though, and that is legitimizing destructive claims to power theologically, that is, in the name of the true God. This would amount to heresy. In the next chapter I will explore the question whether this may have implications for discourse on the Anthropocene.

NOTES

1. Stephen Fowl explores the emergence of idolatry among believers on the basis of the intuition that idolatrous people typically do not regard themselves as such. How, then, do they come to a point where their idolatry is blatantly obvious to others if not to themselves? Fowl suggests that idolatry is a process that slowly, incrementally, misdirects the attention of believers away from the one true God toward other things or people. He thus focuses on the movement from what I call "unsustainable habits of the heart" toward idolatry. See Stephen Fowl, *Idolatry* (Waco: Baylor University Press, 2019).

2. Bob Goudzwaard suggests that ideology may be regarded as "the conduit, track or channel along which idolatry comes to life and moves." He explains the relation between ideology and idolatry in terms of means and ends. The pursuit of a legitimate goal can become an ideology when that end indiscriminately justifies every means toward that end. The obsessions with and blanket justification of such ends legitimate the means so that such means gain power and are turned into idols. An idol thus involves a fixation on particular means or instruments. See Goudzwaard, *Idols of Our Time*, 23.

3. See Institute for Contextual Theology, *The Road to Damascus: Kairos and Conversion* (Johannesburg: Skotaville Publishers, 1989), 11.

4. For this section I used my earlier discussion in *Christianity and the Critique of Consumerism* (Wellington: Bible Media, 2009) as a point of departure. This is updated with some additional literature.

5. See Timothy Gorringe, *The Education of Desire* (London: SCM Press, 2001), 83.

6. See the report *The Consumer Society as an Ethical Challenge* produced by the Church of Norway (1995), 126.

7. *The Consumer Society as an Ethical Challenge*, 127, drawing from *The Road to Damascus* (1989).

8. See, from a Buddhist perspective, David R. Loy, "The Religion of the Market," *Journal of the American Academy of Religion* 65:2 (1997): 275–90 (281), drawing on insights from Norman Brown.

9. Loy, "The Religion of the Market," 276.

10. Ibid., 275.

11. See William Stringfellow, "Money," in *Simpler Living Compassionate Life: A Christian Perspective*, edited by Michael Schut (Denver: Morehouse Group, 1999), 67–72 (68–69).

12. See Serres, *The Incandescent*, 214.

13. See Bob Goudzwaard, "Economie als Afgod," in *Economie als Afgod? Kernvragen van het Conciliar Proces*, edited by Herbert van Erkelends (Kampen: JH Kok, 1989), 21–31.

14. Quoted in Goudzwaard, "Economie als Afgod," 17. The reference is to Andrew Carnegie's *The Gospel of Wealth* (originally published in *The North American Review*, 1889). The booklet is available online from the Carnegie Corporation of New York (2017), https://www.carnegie.org/media/filer_public/0a/e1/0ae166c5-fca3-4adf-82a7-74c0534cd8de/gospel_of_wealth_2017.pdf (quotations on page 2, 15).

15. Hamilton, *Requiem for a Species*, 56.

16. See John B. Cobb (jr), "Liberation Theology and the Global Economy," in *Liberating the Future: God, Mammon and Theology*, edited by Joerg Rieger (Minneapolis: Fortress Press, 1998), 27–42 (32).

17. Cobb, "Liberation Theology and the Global Economy," 32–33.

18. Frank Wagner, *Geld oder Gott? Zur Geldbestimmtheit der kulturellen und religiösen Lebenswelt* (Stuttgart: Ernst Klett Verlag, 1984), 100.

19. Ibid., 12, 142.

20. See Church of Norway, *The Consumer Society as an Ethical Challenge*, 127.

21. Michael Northcott, *A Moral Climate: The Ethics of Global Warming* (Maryknoll: Orbis 2007), 7.

22. Ibid., 39.

23. See also Galtung, *Peace through Peaceful Means*, 203.

24. See Harvey Cox, "The Market as God," *The Atlantic Monthly* 283:3 (1999): 18–23 (18). This early essay was developed into a book with the same title, *The Market as God* (Cambridge: Harvard University Press, 2016).

25. See Cox, *The Market as God*, 8.

26. See Ibid., 16.

27. See Ibid., 9–10.

28. See Ibid., 256.

29. SACC, *Climate Change – A Challenge to the Churches in South Africa*, 44–45.

30. See http://w2.vatican.va/content/francesco/en/apost_exhortations/documents/papa-francesco_esortazione-ap_20131124_evangelii-gaudium.html (accessed January 12, 2019), my emphasis.

31. Messer, "Sin and Salvation," 132.

32. See Evy McDonald, "Spending Money as if Life Really Mattered," in *Simpler Living, Compassionate Life: A Christian Perspective*, edited by Michael Schut (Denver: The Morehouse Group, 1999), 50–66 (59).

33. Ibid., 59–60.

34. See Graham Ward, *Cities of God* (New York: Routledge, 2000), 56.

35. Hans Küng, *Eternal Life?* (London: Collins, 1984), 189.

36. See Sloterdijk, *In the World Interior of Capital*, 231. He describes this addiction to a sudden release of energy as a core component of a globalized energy fascism.

37. Michael Northcott observes that speed produces an altered state of consciousness. It is the form of ecstasy that the technological revolution bestowed on humankind (Milan Kundera). He adds that "This form of ecstasy generates detachment from the violent effects of the ownership and use of the devices of speed on millions who have been killed and seriously injured by them, or from their systemic effects on the ecology of the earth." The car and the speed which it allows gives its owner a sense of control over the physical environment and an unhindered form of freedom. See Northcott, *A Moral Climate*, 218.

38. Ted F. Peters, *Fear, Faith and the Future* (Minneapolis: Augsburg, 1980), 32.

39. Jung Mo Sung, *Desire, Market and Religion* (London: SCM Press, 2007), 44. Sung adds that this does not imply that no sacrifices are being made. Those who become outcasts, the invisible class at the margins of society, are being sacrificed so that the consumer class may enjoy paradise.

40. See Jürgen Moltmann, *God for a Secular Society: The Public Relevance of Theology* (Minneapolis: Fortress, 1999), 88–89.

41. Sallie McFague, *Life Abundant: Rethinking Theology and Economy for a Planet in Peril* (Minneapolis: Fortress, 2001), 93.

42. Mary Grey, *Sacred Longings: The Ecological Spirit and Global Culture* (Minneapolis: Fortress, 2004), 17.

43. Sloterdijk, *In the World Interior of Capital*, 172.

44. Douglas John Hall, *The Cross in our Context: Jesus and the Suffering World* (Minneapolis: Fortress, 2003), 96–97.

45. Segundo Galilea, *Spirituality of Hope* (Maryknoll: Orbis, 1988), 32.

46. See McDaniel's *Living from the Center*, which draws from process theology, Buddhism, and contemplative Christianity and which responds precisely to this sense of emptiness.

47. Gregory Jones, "A Thirst for God or Consumer Spirituality? Cultivating Disciplined Practices of Being Engaged by God," *Modern Theology* 13:1 (1997): 1–26 (18).

48. Jones, "A Thirst for God or Consumer Spirituality?" 18.

49. Ibid., 2.

50. Ibid., 13.

51. John Calvin, *Institutes of the Christian Religion*, edited by John T. McNeill, translated by Ford Lewis Battles (Louisville: Westminster John Knox Press, 1960), 35 (I.1.1).

52. Quoted in Angus, *Facing the Anthropocene*, 111.

53. See Van Hoogstraaten, *Geld en Geest*, 49.

54. Neiman, *Evil in Modern Thought*, 315.

55. Mark Sagoff, "A Theology for Ecomodernism: What Is the Nature We Seek to Save?" *Breakthrough Journal* 5 (2015).

56. Neiman, *Evil in Modern Thought*, 156, in conversation with David Hume.

57. While there is a vast literature on religious diversity that cannot be surveyed here in any way, not many have related this to the Anthropocene. One exception is the brief discussion by Bruno Latour of Jan Assman's work on "translation tables" for the names of gods that were worshipped in different localities. This assumed a rough equality (translatability) between the gods long before the God of Akhenaten and the God of Moses. Assmann adds this provocative statement: "The distinction between true and false religion . . . was unknown to traditional, historically evolved religion and cultures. Here the key differences were those between the sacred and the profane or the pure and the impure. Neglecting an important deity amounted to a far more serious offence than worshipping false gods, the chief concern of secondary religions. In principle, all religions had the same truth-value and it was generally acknowledged that relations of translatability pertained between foreign gods and one's own." See Latour, *Facing* Gaia, 155. The reference is to Jan Assmann's *The Price of Monotheism* (Stanford: Stanford University Press, 2010), 23.

58. See Goudzwaard and Bartholomew, *Beyond the Modern Age*, 33.

59. See Van Hoogstraaten, *Geld en Geest*, 28.

60. Van Hoogstraaten, *Deep Economy*, 3.

61. Van Hoogstraaten, *Geld en Geest*, 38.

62. See Stephen Toulmin's argument in *Cosmopolis: The Hidden Agenda of Modernity* (New York: The Free Press, 1990).

63. See Latour, *Facing Gaia*, 157.

64. Ibid.

65. Hamilton, *Defiant Earth*, 101.

66. Mark Lynas is unashamedly reductionist on this point: "All life is reducible to chemistry—there is nothing more to it than that. No essential life force, no soul, no afterlife." See Lynas, *The God Species*, 5.

67. Northcott, "On Going Gently into the Anthropocene," 25.

68. Harari, *Homo Deus*, 24.

69. Ibid., 258, 259.

70. Lynas, *The God Species*, 29.

71. See also Grinspoon, *Earth in Human Hands*, 214.

72. Sloterdijk, *In the World Interior of Capital*, 209.

73. Neiman, *Evil in Modern Thought*, 109.

74. Ibid., 114.

75. Quoted in Goudzwaard and Bartholomew, *Beyond the Modern Age*, 61.

76. See Hamilton, *Earthmasters*, 111, with reference to positions adopted by Brad Allenby. The reference is to Brad Allenby, "Earth System Engineering and Management," *IEEE Technology and Society Magazine* (2000/2001): 10–24.

77. See again Ellis, "The Planet of No Return: Human Resilience on and Artificial Earth," 43.

78. Ibid., 38.

79. Ellis, "Neither Good Nor Bad."

80. Ibid.

81. Also quoted in Lynas, *The God Species*, 22. The reference is to Brand, *Whole Earth Discipline: An Ecopragmatist Manifesto* (New York et al.: Viking, 2009), epitaph on page 1.

82. See Grinspoon, *Earth in Human Hands*, 165.

83. Northcott, "On Going Gently into the Anthropocene," 32.

84. Haraway, *Staying with the Trouble*, 47.

85. Angus, *Facing the Anthropocene*, 112.

86. Ibid.

87. Ibid.

88. This echoes the words of Psalm 121:1. In secular discourse on the Anthropocene few would dare to add, "My help is from Yahweh who made heaven and earth."

89. Morton, *Hyperobjects*, 201. The reference is to Martin Heidegger, "Nur noch ein Gott kann uns retten," an interview published in *Die Spiegel* (May 1976): 193–219. Morton translates *Gott* with "god" (without the capital).

90. Hamilton, *Requiem for a Species*, 210.

91. Hamilton, *Defiant Earth*, xi.

92. Hamilton, *Requiem for a Species*, 221.

93. Ibid.

94. Ibid., 222.

95. Hamilton, *Earthmasters*, 180.

96. Ibid.

97. See Ernst M. Conradie, "What is the Ecological Significance of God's Transcendence?" in *Nature and Beyond: Transcendence and Immanence in Science and Religion*, edited by Michael Fuller et al. Issues in Science and Theology (Heidelberg: Springer, 2020), 87–99.

98. Hamilton, *Earthmasters*, 183.

99. For a discussion, see Gijsbert van den Brink, "Should we Drop the Fall? On Taking Evil Seriously," in *Strangers and Pilgrims*, edited by Eddy Van der Borght and Paul van Geest (Leiden: Brill, 2011), 761–77.

100. Jung Mo Sung, "Greed, Desire and Theology," in *The Greed Line: Tool for a Just Economy*, edited by Athena Peralta and Rogate Mshana (Geneva: WCC, 2016), 43–54.

101. See Ibid., 53. The reference is to Hinkelammert's *Hacia una critica de la razón mitica: El laberinto de la modernidad* (Mexico: Driada, 2008).

102. See also Cox, *The Market as God*, 21.

103. Sung, "Greed, Desire and Theology," 53.

104. See Joerg Rieger, *Jesus vs. Caesar: For People Tired of Serving the Wrong God* (Nashville: Abingdon, 2018).

105. Ibid., 20.

106. Ibid., 15.

Chapter 8

Heresy in the Anthropocene?

APARTHEID AS A HERESY

The theological legitimation, tacitly or explicitly, of the oppressive political and economic system of apartheid was widely critiqued as a form of heresy.[1] Such a critique indicates that heresies are not focused on relatively trivial matters on the periphery but at the very center of the Christian faith. From a Christian theological perspective, heresy entails a distortion of the truth revealed in Jesus Christ, and something that leads not only to division within the church but also to false witness in the world. This indicates that a point has been reached where a disagreement is no longer merely a matter of opinion, that is, where debate is no longer possible. In a particular context things have become so crucial and divisive that to leave the matter aside would simply not do. The point is that heresy exacerbates violence and often prompts violence—violence against the unity of the church, against human dignity, and against harmony in society.[2]

It should be noted that such a theological critique is an internal one, that is, it is aimed at others within the Christian fold who are endangering the heart of the gospel with distorted views. It cannot be a prophetic critique of what is wrong with the world only. It would not be found in secular literature although the word "heresy" is readily adopted where need be. Once the critique of heresy is made public, it can no longer be regarded as hidden or subconscious but invites theological debate. Such debates often extend over decades if not centuries. However, it is possible to reach a point where debate is no longer possible, where the truth of the gospel is at stake, where there is a need to confess faith and denounce error. Such a point was reached in a "moment of truth" (kairos) when a *status confessionis*[3] was declared in 1982 by the then Dutch Reformed Mission Church (DRMC) in South Africa. The declaration of apartheid as a heresy prompted what was known as a movement toward a confessing church at that time.[4]

This followed upon an appeal from the Lutheran World Federation in Dar-es-Salaam in June 1977 to recognize that the situation in Southern Africa constitutes a *status confessionis*, a Charter by the Alliance of Black Reformed Christians in Southern Africa in 1981 that the theological legitimation of apartheid is a heresy,[5] and a statement to the same effect adopted by the then World Alliance of Reformed Churches in Ottawa in 1982.[6]

In an authoritative essay in *A Moment of Truth* Dirk Smit investigates the background and meaning of the term *status confessionis* in Protestant circles.[7] He notes that there were especially three contexts within which a *status confessionis* was declared in the twentieth century, namely, the church struggle in Germany with regard to loyalty to Hitler and the exclusion of Jews from offices in government and in the church (1934), the ecumenical rejection of the ideology of racism (1977), and of the possession of nuclear weapons (1982). The struggle against apartheid in church and society in South Africa thus constituted a fourth such context.

Smit argues that a *status confessionis* is best understood in a context where Christians or churches have arrived at the judgment that a situation has developed where a moment of truth has arrived within which nothing less than the gospel itself and their most basic Christian confession were at stake so that they feel obliged to witness against that in word and in deed. On this basis he identifies three core aspects of the declaration of a *status confessionis*:

Firstly, the so-called *casus confessionis*, the abnormal situation in which a state of confession is declared is crucial. A *moment of truth* has arrived where the gospel itself has become at stake. In the case of the Belhar Confession, the situation was not merely the oppression and injustices associated with the political system of apartheid. The primary problem addressed in the Belhar Confession, as indicated in the accompanying letter, had to do with the relationship between the then DRMC and the Dutch Reformed Church (DRC) at that time. Although apartheid is not mentioned as such, the issue was the way in which apartheid was propagated by the DRC and legitimized theologically, whether explicitly or tacitly. Moreover, despite numerous exhortations and outcries by the DRMC, the DRC failed to respond appropriately. They apparently did not get the message, believing that their position is orthodox. The mother church did not expect to be taught by the daughter. In this context, the DRMC reached a point where it declared that the integrity of Christian witness in this context was at stake. The only option was to confess the Christian faith as clearly as possible for the sake of the gospel and in the hope that the DRC would recognize what was at stake.

Secondly, it should be noted that the gospel itself was indeed *at stake* in this situation and that neutrality was therefore no longer tenable. This was no mere matter of theological conversation. This is a crucial aspect of Smit's argument. The problem was therefore not merely one of the injustices perpetrated under apartheid which would require prophetic witness in church and society. There was also more at stake than idolatry which would require a call to conversion within the church itself. The primary problem was one of heresy, of false teaching, of thought patterns that became deeply embedded in the church itself, of the failure to recognize and unmask such thought patterns. Such heresy could refer to the explicit theological legitimation of apartheid, but the Belhar Confession went much further than that. In the paragraphs on the various distortions of the gospel that were rejected, it becomes clear that this had to with the failure to embody the visible unity of the church, the various ways in which the ministry of reconciliation in Jesus Christ was obstructed through the legitimation of enforced racial divisions, and the legitimation of various forms of injustice by Christians.

Thirdly, Smit argues that it was indeed the good news of the *gospel* that was at stake. He emphasized that the acceptance of a confession was not aimed to divide the church even further but to plead for reunification. The problem, of course, is that any confessional stance where heresies are denounced would tend to create division even though there is a call to unity on the basis of truth.

In his senior colleague Jaap Durand's contribution to the same volume, the deepest rationale behind the Belhar Confession was clearly articulated. Apartheid was more than an oppressive political system. It became a form of pseudo-gospel and indeed a quasi-soteriology.[8] Many ordinary Christians believed, implicitly but deeply, that separation on the basis of race would help to rescue them from various perceived threats. The way in which they could maintain political power, economic dominance, a particular sociocultural understanding of civilization and safety and security was on the basis of a system of race classification. The argument of the Belhar Confession is that the gospel of reconciliation in Jesus Christ is thwarted on such a basis. It loses any credibility. Thus, there emerged a need to understand the liberating power of the gospel of reconciliation in Jesus Christ across racial and other barriers.

In subsequent debates in South Africa, there were different angles to identify which heresy was at stake: The Belhar Confession focused on the irreconcilability of people—the assumption that races are so different that they can never be reconciled with each other across racial lines. Thus, the reality of not being reconciled is turned into irreconcilability as a principle which is then used to structure society accordingly and becomes legitimized theologically.[9] This negates the gospel of reconciliation in Jesus Christ (a soteriological heresy) and the church as the one body of Christ (an ecclesiological heresy[10]). Another avenue, followed by exponents of black theology such as Allan Boesak,[11] Simon Maimela,[12] and Desmond Tutu,[13] was that apartheid amounted to an anthropological heresy in that it denigrated the dignity of human beings through racism. Racism itself is sin, but the theological legitimation of what amounts to racism is heresy. Other exponents of liberation theology pointed to the failure to understand God's identity and character as the God of the poor and the oppressed who comes to liberate God's people (a *theo*logical heresy).

A *STATUS CONFESSIONIS* ON ECONOMIC INJUSTICES AND ECOLOGICAL DESTRUCTION?

In the context of the World Alliance of Reformed Churches, the possibility of declaring a *status confessionis* in response to the twin issues of ecological degradation and the economic injustices associated with the current global economic order has often been discussed. Typically, the proponents of a *status confessionis* point to the gravity of the situation and look for the strongest possible way in which churches may denounce that.

However, on the basis of the above discussion, some clarity is required on what is meant by "a confessional stance." Why, exactly, does this require a confession? In what way is the integrity of the Christian faith at stake? Injustices in society would call for prophetic witness, not necessarily for Christian confession. Likewise, where such injustices are perpetrated or defended by Christians that would require a call for conversion. Even in the case of idolatry (see above), a state of confession is not required. A state of confession would only be required, at least if the Belhar Confession may serve as a guideline, when the current global economic order is justified theologically by churches or groups of Christians, despite the economic injustices and ecological destruction associated with that. Such theological justification may be explicit through official ecclesial resolutions or may be evident from the teaching and ministries of various Christian groups. Even if such a theological justification may be clearly identified, a *status confessionis* would only be appropriate after many years of Christian witness and attempts to help those propagating heresy to see the demands of the gospel more clearly. A *status confessionis* could only be declared when other avenues have proved fruitless, when ongoing dialogue is no longer possible, when the urgency of the situation demands it, and when confessing one's faith as clearly as possible proves to be the only way forward. As many have argued, it requires a sense of prophetic discernment to know when that moment of truth has arrived—not too early and not too late.

Since it was not at all clear that such a situation had been reached by 1997,[14] the Debrecen assembly of the former World Alliance of Reformed Churches, rightly in my opinion, opted to declare what was called a *processus confessionis*, that is, a "process of recognition, education, confession and action regarding economic injustice and ecological destruction."[15]

A consultation on the impact of globalization in Central and Eastern Europe was held in Budapest, June 2001, and organized by the World Council of Churches, the World Alliance of Reformed Churches, and the Lutheran World Federation in conjunction with the Conference of European Churches. The message from this consultation was, "Serve God, not Mammon!" It raised the following challenging question: "Will the churches have the courage to engage with the 'values' of a profit-oriented way of life as a matter of faith, or will they withdraw into the 'private' sphere?" It suggested, "This is the question our churches must answer or lose their very soul." It asked churches to help their members to rediscover the Christian virtues of self-restraint and asceticism and to "propagate these values in their societies as a way of countering individualism and consumerism, and as an alternative foundation for economic and social development."[16] This clearly indicates a shift from a prophetic critique of injustices to a prophetic critique of idolatry.

At its 24th General Council in Accra, Ghana, July 30–August 13, 2004, the World Alliance of Reformed Churches offered the following statements regarding economic injustice and ecological destruction in what became known as the Accra Confession:

> 16. Speaking from our Reformed tradition and having read the signs of the times, the General Council of the World Alliance of Reformed Churches affirms that global economic justice is essential to the integrity of our faith in God and our discipleship as Christians. We believe that the integrity of our faith is at stake if we remain silent or refuse to act in the face of the current system of neoliberal economic globalization and therefore we confess before God and one another.
>
> 19. Therefore, we reject the current world economic order imposed by global neoliberal capitalism and any other economic system, including absolute planned economies, which defy God's covenant by excluding the poor, the vulnerable and the whole of creation from the fullness of life. We reject any claim of economic, political, and military empire which subverts God's sovereignty over life and acts contrary to God's just rule.
>
> 21. Therefore we reject the culture of rampant consumerism and the competitive greed and selfishness of the neoliberal global market system, or any other system, which claims there is no alternative.
>
> 22. We believe that any economy of the household of life, given to us by God's covenant to sustain life, is accountable to God. We believe the economy exists to serve the dignity and well-being of people in community, within the bounds of the sustainability of creation. We believe that human beings are called to choose God over Mammon and that confessing our faith is an act of obedience.

> 23. Therefore we reject the unregulated accumulation of wealth and limitless growth that has already cost the lives of millions and destroyed much of God's creation.
>
> 25. Therefore we reject any ideology or economic regime that puts profits before people, does not care for all creation, and privatizes those gifts of God meant for all. We reject any teaching which justifies those who support, or fail to resist, such an ideology in the name of the gospel.[17]

The language used in this statement indicates that more than economic injustices and ecological degradation were seen to be at stake. The document also speaks of "the integrity of our faith in God and our discipleship as Christians." It calls for a critique of ideology, especially the ideology of consumerism, and seeks to unmask various form of idolatry. In sections 9 and 10, it observes that the current crisis is "directly related to the development of neoliberal economic globalization" and that "This is an ideology that claims to be without alternative, demanding an endless flow of sacrifices from the poor and creation. It makes the false promise that it can save the world through the creation of wealth and prosperity, claiming sovereignty over life and demanding total allegiance, which amounts to idolatry." In section 25 quoted above, the document also critiques the theological legitimation of this ideology in the name of the gospel.

However, it is not clear from this document where and how such theological legitimation may be found. It hangs in the air, as it were. Compared to the immediate context within which the Belhar Confession emerged (the relationship between the DRC in South Africa and the former DRMC), there is scarcely any reference to sustained exhortations in dialogue with those who perpetrate such heresies. As a result, there remains a certain vagueness in the document—not on economic injustices and ecological destruction, not on the neoliberal economic order, not on the dangers of idolatry, but on the theological justification of such distortions of the gospel. This begs the question: to whom is the Accra Confession really addressed? To those in positions of power in institutions that govern the global economy (if so how are they expected to respond)? To the members of reformed churches for the sake of self-examination (on sins and idolatries in their own midst) and a call for conversion, and subsequently for education, reflection, confession, and action? Or to address those who propagate some form of heresy, whether implicitly or explicitly (but where may that be found)? Is this merely a way of prioritizing global issues on ecclesial agendas? Although the background to the Accra Confession is quite clear—namely, with reference to wider and narrower ecumenical discourse on the global economy—it is not equally clear to whom it is addressed and with what expected outcomes.

THE PROSPERITY GOSPEL AS HERESY?

On the basis of the preceding analysis, one may conclude that there are multiple forms of idolatry at play not only in society but also in the midst of Christian communities in the context of globalized consumerism. As I argued above, the presence of such forms of idolatry would not by itself warrant the proclamation of a *status confessionis*. The question is whether such idolatry is also justified in terms of Christian teaching and whether it is somehow defended theologically (e.g., in ecclesial resolutions). The clearest and crudest examples of such a theological justification of the culture of consumerism may be found in contexts where the prosperity gospel is being propagated through Christian preaching and teaching but scarcely through ecclesial resolutions. Could this be regarded as a form of heresy?

I will refrain from engaging in a detailed investigation of the nature and impact of prosperity gospel. This is a topic that would require a survey in its own right, based on empirical data from different countries. It is at least clear that this form of Christianity has spread from the United States and elsewhere in industrialized countries to the African continent, especially among new Pentecostal-Charismatic Churches (PCCs). It is evidently thriving in South Africa and from what my students tell me certainly also on the Cape Flats. It is likely that such a portrayal of the gospel in terms of success, prosperity, and wealth will become increasingly influential in the South African context, especially given the current role of paid television channels, and the broadcasting of religious programs derived from American sources and styles.

There is a rapidly expanding corpus of publications on the prosperity gospel, with diverging assessments, clearly influenced by particular contexts. Two helpful review articles may suffice here: In an informative review based on case studies in Nigeria, Brigitte Meyer describes the emphasis on appearance and status symbols among the leaders of such new Pentecostal churches: "The leaders who dress in the latest (African) fashion, drive nothing less than a Mercedes Benz, participate in the global Pentecostal jet-set, broadcast the message through flashy TV and radio programs, and preach the Prosperity Gospel to their deprived and hitherto hopeless born-again followers at home and in the diaspora."[18] Meyer argues that the prosperity gospel is the main attraction of PCCs since people's felt needs are apparently met, but, as the promise of prosperity fails to materialize among ordinary believers, also the main weakness of the prosperity gospel. She also notes that crude forms of consumerism are denounced in such circles. Prosperity is the blessing bestowed by the Holy Spirit upon those who use money wisely and who refrain from alcohol and sexual promiscuity. The prospect of prosperity is therefore dependent upon deliverance from seductive powers.[19]

In an excellent recent overview, Andreas Heuser recognizes that it is impossible to generalize on the socioeconomic impact of the prosperity gospel given urban/rural divides, class differences among proponents and excesses. He states, "A township-based ethics of survival is different from an urban and middle-class 'progressive' Pentecostalism; while business management oriented churches have aspirations to transform society, the socio-economic horizon of churches entertaining a strong sacred secrecy around prosperity rather remains short-term and confined to internal dynamics."[20]

Like most other forms of heresy, the prosperity gospel flourishes on an element of truth, but also systematically distorts that. The element of truth at stake here is that of God's blessings—including very concrete and material blessings ascribed to God's providence—such as enough rain on time, today and tomorrow's bread on the table, protection on the roads, success with one's studies, deliverance in times of crisis, and enough income to live from. In certain contexts money can indeed be the way in which God would bless people. For those trapped in a culture of poverty refraining from alcohol and drug abuse, visiting prostitutes, borrowing money and gambling, doing honest hard work, spending money frugally, and a commitment to the needs of one's family may well lead to increasing prosperity in a material sense of the word. Who would wish to argue that this is not a concrete sign of God's blessings?

Moreover, the prosperity gospel may easily be employed to legitimize a sense of "upward social mobility." There are many people, for example, on the Cape Flats, who are entering the (lower) middle class in South Africa for the first time. They realize that they themselves may not have received a good education, but that there are opportunities available for their children. If they can escape from the temptations of a culture of poverty, with some hard work and dedication, they can make it in life. They may soon be able to live in a suburban flat or house of their own, buy a car, and perhaps become part of the consumer class. Pastors in such areas typically support such a sense of upward social mobility. They emphasize the role of talents and opportunities for education and training and speak of grabbing such possibilities as being God's will. Such pastors often study at postgraduate level at UWC in order to set an example for their families and communities. Although the lifestyles of the consumer class are not sustainable for all humans on Earth, this theological undergirding for a sense of upward social mobility is again quite understandable, to say the least. It inspires church members within their local communities to aim higher, to believe in themselves, and therefore to work harder so that they can reach the top (whatever that may mean). In short, it seeks to cultivate a sense of hope in communities where hopelessness amid poverty, gangsterism, prostitution, drug abuse, and so forth—after centuries of conquest, slavery, and serfdom—abounds.

However, the prosperity gospel may also be used to encourage overt forms of affluence or some religious version of a "quick fix." In many cases, the underlying assumption is that, if you give your best for the Lord (and for the coffers of the local congregation[21]), you will receive rich blessings from God. Thus, such blessings then become signs of the authenticity of one's faith (and of the pastor's faith). There is something wrong with your faith if you do not receive such blessings. In such contexts, the prosperity gospel may well become heretical.

In my view, the problem related to the prosperity gospel is that it typically speaks about God's blessings outside of the tension between the fall of God's good creation and the redemption of creation. The Christian gospel speaks about the redemption of creation from sin and evil so that its inherent goodness and integrity can be appreciated anew. The narrative of human culture, including the role played by money, buying, and consuming, is situated within this drama of God's work on Earth. This is not a drama of redemption *from* creation but of the redemption *of* creation. Christians therefore own and spend money, buy and consume within this tension—in the midst of the decay of a consumer culture, but also on the basis of the redemption in which Christians may already share, but which has not yet arrived fully.

The prosperity gospel, the gospel of health and wealth, may therefore indeed be regarded as a distortion of an understanding of God's providence. Since the prosperity gospel is propagated most overtly among Pentecostal churches, this creates an ecumenical problem, especially if the critique comes from established churches or ecumenical networks. It may amount to nothing but jealousy from the side of "old money" aligned with classic forms of capitalism regarding the "new money" aligned with neoliberal forms of capitalism. Insofar as this critique is aimed at Pentecostal and other independent churches, it may itself be suspect. The way forward in this regard is to investigate the far-reaching commercializing of churches, of the Christian faith, and of the gospel itself within a consumer society—to address the needs of consumers of religious "products."[22] The danger here is that theological legitimation and ecclesial sanctification may be offered in this way with regard to the lifestyles, values, and tacit assumptions of those Christians in the consumer class while being eagerly critical of the upward social mobility among those who see the need for "development." In short, if the lifestyles of the consumer class are not sustainable and cannot be copied by all, then consumer Christianity can only exacerbate the problem, both by legitimizing the lifestyles of those who have been in the consumer class for decades and of those seeking to join the consumer class.

AND THE ANTHROPOCENE?

What makes the portrayal of apartheid as a heresy intriguing is that this also had some secular appeal, most notably in the critique that apartheid presupposed the fundamental irreconcilability of people in terms of race. Can this also be true of any critique of heresy in the Anthropocene? This is not immediately obvious (to me).

It is indeed far from clear whether heresy is an appropriate angle for a diagnosis of what is wrong with the world in the Anthropocene. In most cases, the critique of injustice, oppression, ideology, and idolatry would suffice. Injustice, structural violence, and ideology rightly focus on ethical matters, but idolatry and heresy are best understood in doctrinal terms. Whereas idolatry focuses on what it is that people put their trust in, heresy is at least also about ways of seeing the world. To suppose or suggest that free-market industrialized capitalism is supported by Christianity and can be defended in the name of the God of Christianity, as is typically presumed in right-wing evangelicalism, may indeed be regarded blasphemous, but not yet as heretical.[23] From a Christian perspective it is best to relate heresy either to the soteriological heart of the Christian gospel, to the Christological core of the Christian faith, and to the Trinitarian width of the Christian creed.

Let me offer a few pointers to where the dominant ways of seeing the world in the Anthropocene may have gone radically wrong, at least from a Christian perspective. Since heresy is best understood as an intra-Christian critique, such ways of seeing the world elsewhere in society should concern but need not surprise Christian communities. However, where this infiltrates an understanding of the Christian faith, it leads to radical distortions. If defended theologically, heresy looms.

a) The Christian confession holds that God the Creator whom Christians have best encountered in Jesus the Christ is the ultimate Source of energy. Where this Source is denied, energy is treated as a "natural resource," available for human exploitation. While energy is abundant, the Industrial Revolution and the subsequent "great acceleration" rely on a frantic scramble for fossil fuels as sources of energy, first coal and then oil. This has led to radical distortions in the economy, best understood as governing the household of God (*oikos* + *nomoi*).

Several theological critics have recognized what has gone wrong here. Three examples may suffice:[24]

In his book *God the Economist* (1989), Douglas Meeks observes that there has been an eclipse of God's household (*oikos*) in the neoliberal notion of the self-regulating market.[25] As a result, the notion of livelihood within the household of God became truncated and was eventually replaced by a notion of scarcity defined by the logic of exchange within the market. The logic of the market creates artificial scarcity and satiation. He observes that both scarcity and satiation deaden the spirit and impedes life with the other. He contrasts this with satiation in a consumer culture and a consumer-oriented church:

> The effect of the practice of scarcity based market-logic is satiation. Both scarcity and satiation deaden the spirit and impede communal life with the other. Artificial scarcity spawns the lottery culture: The others may not make it, but I may. Satiation slakes the thirst for righteousness. Scarcity and satiation leave only one possibility for the distribution of what is necessary for life and life abundant: the logic of exchange. The genuinely other cannot appear in this logic, nor can the reality of grace.[26]

Meeks argues that, subsequently, there emerged what he calls "God concepts" within the logic of the market society that justify the prevailing view of reality behind the market logic, including exclusive property rights, an accumulation of capital, the need for wage labor and human acquisitiveness.[27] While desire, hunger, wants, and needs spelled trouble for humans in premodern times, needs are seen to be the impetus to vitality, productivity, and success in modernity.[28] The human project is seen as the progressive, mutual meeting of human needs.[29] An acquisitive orientation forms the presupposition and guarantees the unlimited growth of the economy. On this basis God is portrayed as one who has no needs (the aseity of God), who has the ability to decide without any constraints (God's sovereignty) and who is infinite (God's infinity). In a consumer society these three beliefs are functionally replaced by the notion that "the consumer is king." Once a human need can be identified, there is an immediate justification for meeting such a need, irrespective of how extravagant that may be. Accordingly, consumers are portrayed as independent, sovereign, and insatiable.

Meeks contrasts this with the Christian portrayal of the Triune God as "the Economist" and emphasizes the fullness (*pleroma*) of God's blessings and the gifts of the Holy Spirit over against such a notion of scarcity and human needs. As in the exodus narrative of God providing abundant manna in the wilderness, the pretense of scarcity as the foundation for economics cannot be tolerated.[30] The logic of God's economy is not based on scarcity but on plenitude and overwhelming generosity: "The Lord is my shepherd, I shall not want" (Ps. 23). Thus, the management of the household of God becomes a matter of liturgical praise for God's abundant grace and the just distribution of God's blessings following a Eucharistic logic.[31]

In the school of "Radical Orthodoxy," Graham Ward has argued that unlimited consumer desire is a manifestation of what he calls an "economy of lack" at the heart of modernity.[32] He analyzes the social atomism of contemporary urban existence and the fiction of the self-sufficient consumer as an extension of the self-isolated Cartesian subject. As a result, relationships with others are reduced to acts of narcissistic consumption, desire to libido, and intimacy to orgasm. The consumer seeks to break through such self-enclosure through desiring something that is "other," but such desire cannot be fulfilled and therefore engenders obsessive and possessive longing. The economy of lack presupposes distance and difference between the self and the other which is to be consumed. Ward explores such desire on the basis of psychoanalytical theory, following the intuition that libidinal desire is inseparable from consumerist desire. He seeks to counter such a notion of desire resulting in social atomism on the basis of a Barthian divine call from beyond the self toward participation—in the life of the Triune God, in the body of Christ, in the Eucharist, and in an analogical network of social relationships. The doctrine of participation that Ward portrays is one which allows for desires, deepens longing, celebrates joy, and plunges all of that into a more sublime sense of wonder which he describes in terms of Augustine's notion of joy in God.[33] The self is here portrayed as not self-enclosed but in a relational openness to the world and to God. This call from beyond the self is structurally similar to the vision of a heteropolis luring us beyond the current consumerist polis toward the "city of God" (Augustine).

In his book *Divine Economy*, Stephen Long offers a critique of a "heretical ontology of lack." He argues that the basic assumption of modern economics is that there is a scarcity of resources. The problem to which the discipline of economics therefore responds is the allocation and management of scarce resources amid competing needs. The individual consumer confronts finite commodities with a desire grounded in infinite freedom. To satisfy a desire with one commodity is to forego all others so that the experience of scarcity is inevitable. The infinite desire of consumers exacerbates such scarcity on a finite planet. Daily life in a consumer society trains consumers in countless way to see the world in terms of a lack of sufficient resources. Thus, the human desire for the infinite is reduced to an insatiable desire for the finite. As a result, consumers create an alternative hierarchy that leaves them unable to see the ultimate *telos* of human existence, namely, fellowship with God.[34] For Long, God is not defined by lack but by plenitude. The grateful recognition of God's grace, the abundant ultimate source of energy, would allow for joy, delight, playfulness, and an ability to give things away because they are not really possessed in any case. An ontology of grace and gratitude could thus provide a point of entry to counter a heretical ontology of lack.

In a consumer society such a notion of *pleroma* is no longer regarded as a gift but becomes a demand. While a sense of *pleroma* evokes a response of gratitude and frugality, the massification of surplus consumer products, based on fossil fuels as source of energy, evokes a sense of frivolous and wasteful consumption. As Peter Sloterdijk observes, "Mass frivolity is the psychosemantic agent of consumerism." He adds that "the ban of wastefulness has been replaced by the ban on frugality, expressed in the perpetual appeals to encourage domestic demand."[35]

b) Whose Anthropocene is it anyway? How is this "Anthropos" after whom the new geological epoch may well be named construed across divides of geography, race, class, gender, culture, and religion? From the above discussion, it is evident how contested this Anthropos is and how difficult it is to balance both a decentering and a recentering of the place of humanity in God's household.

For Christians, such debates cannot but evoke a reminder of Pontius Pilate's ironic exclamation, "Ecce, homo!" The grandeur and misery of the human condition is indeed well captured by the figure of the crucified Messiah, the all too visible image of the invisible God. The Christian confession that "Jesus Christ is Lord" is often understood as a glorification of Jesus's divine status (Homo Deus!). This can at best be a half-truth given the inverse, namely, that the "Lord" is this crucified one. This cannot but challenge any colonial and patriarchal notions of lordship and redefines, indeed radically undermines what being "lord" (e.g., a landlord) may mean.

One may observe that the classic Christological debates with all their accompanying heresies are being revisited in debates on the Anthropocene but now with the assumption that what applies to Christ as "fully human," indeed "fully animal," and "fully divine" applies to all humans (but humans only?). This extension of divine status to all human beings, presumably in order to honor human dignity but often with Promethean aspirations, cannot but yield confusion, blurring the distinction between creature and Creator and the solidarity of humans with their evolutionary origins. Where the uniqueness of Christ is sacrificed, so that there are many other ways up the mountain (many messiahs), all followers of Jesus Christ need to become like him, also as far as divine status is concerned. Alternatively, the distinction between Creator and creature can be regarded as alienating, leading to a desacralizing of what is natural, so that the blurring of this distinction is praised. This would yield a form of pantheism (often associated with environmentalism) that would still need to come to terms with the presence of evil. In response, Christian commentators have often warned against "Worshipping the Creation Rather than the Creator."[36]

Such confusion is aggravated by Athanasian views holding that in Christ God became human so that humans can be made divine (*theopoiēsis*).[37] Latter-day followers of Athanasius, the Nicene champion of orthodoxy, would need to guard against heretical distortions of their soteriological views on *theosis* in the Anthropocene. With discourse on the God species, it seems that Athanasius has gained quite a secular following! As was the case in Patristic theology, a heresy may become manifest in the person of Christ, but the reason why that heresy had to be denounced was and remains of a soteriological nature. It is our salvation that is at stake. This is also why heresies matter in secular contexts, namely, to expose false remedies and false saviors.

The additional danger is that any perception of humans as the God species may revert to a Manichean disdain for being embodied, where humans are regarded as nothing but dust, a worm and not a human, with the resulting hope to transcend carbon-based forms of life toward silicon-based forms of intelligence. Why carbon (a core ingredient in all forms of life) is treated as dirty while silicon (found in desert sand) is not, is of course not clarified. If humans are degraded in this way, it should not surprise us that all humans (not just Jesus Christ) are regarded as being without sin since they are victims of circumstances beyond their control so that, if anything, God is to be blamed for whatever they may do wrong.

c) The Anthropocene presents an odd interplay between immense power and powerlessness. Consider, on the one hand, the cumulative impact of the global economy to the extent that humans have become a geological force of nature, the unleashing of planetary forces far beyond human control in which the "awakened monster" may "crush" its tormentors, and the Promethean aspirations of geoengineering that suggest that human engineers may be able to "turn the thermostat" with their fingertips in order to adjust global surface temperatures according to cultural wishes and commands. Consider then, on the other hand, the apparent inability, indeed the feeble impotence, of societies, politicians, economists, citizens to alter the energy basis of the global economy timeously in order to avoid catastrophic climate change. Splitting something as small as an atom may unleash unprecedented energy, while global superpowers and corporate giants in the fossil fuel industry are splitting hairs over appropriate carbon taxes and energy subsidies.

The Christian confession of faith in the Holy Spirit, the Spirit of the crucified Christ, is that this Spirit is the "Giver of Life," who brings forth life from God's abundant love and who brings forth new life amid barrenness, destruction, desolation, even death. For Christians it is not money that makes the world go round but the Spirit of Love, working through vulnerability and manifested in the powerlessness of the cross. It expresses the counterintuitive hope, based on the promises of God and the partial fulfillment of these promises in history, that the God who seem to have abandoned God's "beloved Son" has not abandoned the living creatures beloved by God, including the many victims (human and otherwise) of history. This requires discernment, to read the signs of the time and to detect the movement of God's Spirit in history.[38]

Pneumatological forms of idolatry would put trust in other forms of power—military, technological, financial, political, managerial, physical, even the power of positive thinking. These would constitute a failure of nerve, not believing in the power of God's love that surpasses all understanding. Pneumatological forms of heresy have perhaps been scarce in the history of Christianity, except for the tacit denial that the Spirit of God is indeed the Spirit of Christ. My sense is that such heresy looms in the Anthropocene in the abandonment of hope. This is perhaps related to sin as sloth, an acquiescence in the status quo, giving up hope that things could still change. This yields eschatological heresies such as escapism and nihilism. Amid the looming destruction of the Anthropocene it accepts that God's Spirit has abandoned the Earth. All we can then do is to try and save ourselves and our loved ones—to put up some fight even if this is inevitably futile.

Such a retrieval of Christian hope will remain facile if the future that the Anthropocene holds is not confronted. Christian hope is not the same as optimism, as it is not based on an extrapolation of current trajectories but on the radical promises of God. This is symbolized by the biblical image of the "valley of dry bones" in Ezekiel 37. New life can spring forth where there seems to be no hope. Nevertheless, blind hope is as risky as blind faith is misleading. Once the future is confronted, the attractiveness of eschatological heresies becomes evident. This calls for further reflection on where the world is heading toward. I will proceed to offer some reflections in this regard in the form of an apocalyptic theological postscript.

NOTES

1. See, especially, De Gruchy and Villa-Vicencio, *Apartheid Is a Heresy.*
2. As captured by Nico J. Smith, "Apartheid in South Africa as a Sin and Heresy: Some of its Roots and Fruits," in *New Faces of Africa: Essays in Honour of Ben (Barend Jacobus) Marais*, edited by Jan W. Hofmeyr and Willem S. Vorster (Pretoria: Unisa, 1984), 143–52.
3. There is extensive discourse on declaring a *status confessionis* in reformed literature. See especially Smit, "What Does *Status Confessionis* Mean?"; also Ulrich Möller, "*Status Confessionis*? Confessing Our Faith in the Context of Economic Justice," *Reformed World* 46 (1996): 138–44; and Ernst M. Conradie, "Globalisation, Consumerism and the Call for a *Status Confessionis*," in *Globalisation Volume II: Global Crisis, Global Challenge, Global Faith – An Ongoing Response to the Accra Confession*, edited by Allan A. Boesak an Len Hansen (Stellenbosch: SUN Press, 2010), 53–76.

4. This connection is made explicit in an essay by John W. de Gruchy, "Towards a Confessing Church: The Implications of a Heresy," in *Apartheid is a Heresy*, edited by John W. de Gruchy and Charles Villa-Vicencio (Cape Town: David Philip, 1983), 75–93.

5. This Charter draws on the classic distinction between a true and a false church, recognizes the theological difficulties in making such a distinction since the empirical church always includes a mixture of both, but then concludes: "We, as members of ABRECSA, unequivocally declare that apartheid is a sin, and that the moral and theological justification of it is a travesty of the Gospel, a betrayal of the Reformed tradition, and a heresy." See the appendix in De Gruchy and Villa-Vicencio, *Apartheid is a Heresy*, 162.

6. See the appendices in Ibid., 168–73.

7. See Smit "What Does *Status Confessionis* Mean?" The discussion below draws verbatim from my essay "Globalisation, Consumerism and the Call for a *Status Confessionis*."

8. See Durand, "A Confession – Was it Really Necessary?".

9. See, for example, Jaap Durand, "Die Kerk se Profetiese Roeping," in *Perspektief op die Ope Brief*, edited by David J. Bosch, Adrio König and Willem Nicol (Kaapstad: Human & Rousseau, 1982), 66–85 (69); also H. Russel Botman, "Narrative Challenges in a Situation of Transition," in *To Remember and to Heal: Theological and Psychological Reflections on Truth and Reconciliation*, edited by H. Russel Botman and Robin M. Petersen (Cape Town: Human and Rousseau, 1996), 37–44.

10. This is the angle taken by David Bosch in an essay entitled "Nothing but a Heresy," in *Apartheid is a Heresy* edited by John W. de Gruchy and Charles Villa-Vicencio (Cape Town: David Philip, 1983), 24–37.

11. See Allan Boesak, "He Made us All, but . . .," in *Apartheid is a Heresy*, edited by John W. de Gruchy and Charles Villa-Vicencio (Cape Town: David Philip, 1983), 1–9.

12. See Simon Maimela, "An Anthropological Heresy: A Critique of White Theology," in *Apartheid is a Heresy*, edited by John W. de Gruchy and Charles Villa-Vicencio (Cape Town: David Philip, 1983), 48–58.

13. See among many of his other writings on human dignity, Desmond Tutu, "Christianity and Apartheid," in *Apartheid is a Heresy*, edited by John W. de Gruchy and Charles Villa-Vicencio (Cape Town: David Philip, 1983), 39–47.

14. See the excellent article by Ulrich Möller, entitled "*Status confessionis*?" Möller argues that the impact of the current global economy should not be regarded as a *casus confessionis*. Although there can be no doubt about the ethical challenges that global capitalism poses, a *status confessionis* can only emerge in a situation where the gospel itself is at stake and where it is necessary to maintain the unity of a church threatened by heresy. He suggests that the declaration of a *status confessionis* can only hamper communication and impair the openness that the confessing process requires.

15. See the document entitled *Processus Confessionis: Process of Recognition, Education, Confession and Action Regarding Economic Injustice and Ecological Destruction* (1997), Background Papers No. 1 of the World Alliance of Reformed Churches.

16. Quoted in *Alternative Globalization Addressing Peoples and Earth* (WCC, 2005), 49. See this document for a number of other ecumenical consultations on globalization as an issue of faith.

17. For the Accra Confession, see http://wcrc.ch/accra/the-accra-confession (accessed September 2, 2019).

18. Brigitte Meyer, "Christianity in Africa: From African Independent to Pentecostal-Charismatic Churches," *Annual Review of Anthropology* 33 (2004): 447–74 (448).

19. Ibid., 460.

20. Andreas Heuser, "Charting African Prosperity Gospel Economies," *HTS Theological Studies* 72:4 (2016): 1–9 (8).

21. For a critique of such abuse from within the context of "Coloured Independent Churches," see Brian Herbert, "A Conceptual Analysis of Abusive Charismatic Leadership," M.Phil. mini-thesis, University of the Western Cape (2008).

22. See the fifth point of entry for a critique of consumerism as discussed in my *Christianity and a Critique of Consumerism*, namely, that the consumer society encourages commercialized cultural and religious practices.

23. See De Gruchy, *The End is Not Yet*, 49.

24. I am drawing here from paragraphs in the final chapter of my *Christianity and a Critique of Consumerism.*

25. M. Douglas Meeks, *God the Economist: The Doctrine of God and Political Economy* (Minneapolis: Fortress, 1989).

26. M. Douglas Meeks, "The Future of Theology in a Commodity Society," in *The Future of Theology: Essays in Honor of Jürgen Moltmann*, edited by Miroslav Volf, Carmen Krieg and Thomas Kucharz (Grand Rapids: Eerdmans, 1996), 253–66 (263).

27. See Meeks, *God the Economist*, 58.

28. Ibid., 159.

29. Ibid., 160.

30. Ibid., 175.

31. See also Meeks, "The Future of Theology in a Commodity Society," 263.

32. Ward, *Cities of God*, 172.

33. Ibid., 173.

34. See D. Stephen Long, *Divine Economy: Theology and the Market* (London: Routledge, 2000), 143–71.

35. Sloterdijk, *In the World Interior of Capital*, 228. Sloterdijk is correct to recognize that notions of the luxurious and superfluous are based on theological judgments but misses a recognition of God's *pleroma*. He states that "everything superfluous could only be displeasing to God" (228).

36. See Sagoff, "A Theology for Ecomodernism."

37. For a discussion in the context of Christian ecotheology, see Denis Edwards, *Deep Incarnation: God's Redemptive Suffering with Creatures* (Maryknoll: Orbis, 2019), 67–69.

38. See Ernst M. Conradie, "What Makes the World Go Round? Some Reformed Perspectives on Pneumatology and Ecology," *Journal of Reformed Theology* 6 (2012): 294–305.

Chapter 9

An Apocalyptic Theological Postscript

TELLING THE STORY IN THE HERE AND NOW

As many secular scholars would agree, the Anthropocene requires a retelling of the story of the past, present, and future. Since time immemorial stories of origins have told us humans where we come from, who we are, how we are related to the Earth and to other animals, what role(s) we play and should play in the bigger scheme of things, and how we should cope with challenges from the inside and the outside. Such stories often give humans a privileged place at the center of a cosmos with ordered relations and legitimize the role of humans, for example, as gardener, steward, ruler, explorer, warrior, conqueror, inventor, or destroyer. All over the world such roles have been defined and redefined when narratives came into contact with each other and when new insights require a retelling of that story. Environmentalism is regarded as a secular form of faith precisely because it raises concerns over humanity's ultimate place and purpose in the world.[1] When stories told about ecological destruction become connected with each other, mythos has implications for an appropriate ethos, praxis, and politics.

The recognition of the Anthropocene, that Earth systems are being reshaped by the human species, clearly requires a revised version of this story.[2] The reason for that is that the Anthropocene puts humans back in the center of planetary history whereas scientific developments in cosmology (Copernicus), geology, evolutionary biology (Darwin), and astrophysics have progressively decentered the emergence of humans in the history of the planet and the universe. This task can no longer be addressed separately by the so-called humanities and the so-called natural sciences. Michel Serres, for one, recognizes the need for philosophy not only to focus on the "black hole" of social relations (the legacy of the slave trade) but to rejoin its own planetary habitat and its grand narrative.[3]

There is a need to explain how we reached this point in planetary history but also a need to guard against distortive versions of such a story. One example is the triumph of consumerism (and boredom), as the end goal of capitalism so that communism is best understood as "a stage on the way to consumerism,"[4] declaring that "the end of history" has arrived. Another example is the narrative that allows policy makers to continue with business as usual with a few added environmental cautions. Indeed, "The delay-climate-action-and-make-nature-pay-later story is not a wise one to tell ourselves."[5] We need a compelling story to guide our collective human actions around mitigation and adaption in a changing climate. Beyond that, how is time's arrow to be understood?[6] What kind of geological epoch or era will follow the Anthropocene? Will some new balance return where humans are no longer the main agents of geological change? Will there still be humans present?[7]

Where can such a story be found? This is a triple question: Do the integration of Earth systems also require an integrated version of the story? Can such a story be found? Is there a proper place where such a story can be told? In the Anthropocene the category of place itself has become contested: whose place is in anyway?[8] Added to this is the problem that a holistic perspective (Earth system science?) is not only impossible but is also arrogant. As Bruno Latour rightly observes, "One is never as provincial as when one claims to have a 'global' view."[9] And: "he who looks at the earth as a Globe always sees himself as a God."[10] If we tell the story from somewhere, we cannot tell the whole story, not least because the story is still ongoing. We tell it only in fragments, necessarily with the distortions that any particular perspective brings.[11] In doing so, we still need to allow for a sense of mystery, for that which cannot be made to fit within a story, for the kind of truths that really matter.

How, then, is the narrative of the Anthropocene to be told? All disciplines reflecting on the Anthropocene ultimately need to reflect not only on its starting point and subsequent history but also on its end(s). Can one avoid grand narratives (or at least mega-narratives) in this regard? One may speak with Holmes Rolston in terms of three big bangs: the evolution of the cosmos, of life, and of consciousness.[12] Michel Serres acknowledges that grand narratives cannot be avoided and believes that the Enlightenment universalism has been broadened four times with the cosmic big bang, the cooling of our planet, the appearance of RNA and of *Homo sapiens*.[13] But how is a new grand narrative of history in stages that merely perpetuate the possibility of domination to be avoided?[14] Christophe Bonneuil observes that such narratives attribute a certain value to the state of things at the beginning and at the end of the story, select a focus and a framing that highlights some actors while leaving others in the shadow, put time into sequences, pinpointing certain turning points while downplaying others, and constitute a dramaturgy with implicit or explicit moral lessons.[15] He outlines four grand narratives of the Anthropocene:

The dominant grand narrative is the modernist one popular in scientific cultures where "Man" moves from environmental obliviousness to environmental consciousness, from "Man" equaling nature's power to "Man" repairing nature through the role of science. The second, utopian (or ecomodernist) grand narrative celebrates the end of nature and welcomes the advent of the "good Anthropocene" and thus radicalizes the modernist version of progress and Enlightenment. The third, eco-catastrophist narrative warns of the collapse of industrial civilization by depicting limits, tipping points, collapse, violence, and wars. This narrative rejects the faith that greener technologies can save the planet: "It argues for the urgent need to radically change the dominant ways of living, consuming and producing, and rejects the belief in technological fixes that would save the planet within the frame of an unchanged socio-economic system."[16] Science and technology alone cannot save the planet; only social innovations emerging from a dynamic civil society could. The fourth, eco-Marxist narrative focuses on capitalism's inability to ensure sustainability due to the processes of dispossession and commodification associated with capitalist expansion and imperial domination. This narrative privileges the term Capitalocene rather than Anthropocene. He notes that one may add similar ecofeminist and subaltern narratives that relate domination in the name of differences of gender and caste to the degrading of the Earth.[17]

Indeed, how is the story of the advent of the Anthropocene to be told? Can such a story be told, if the Anthropocene is indeed unprecedented in human history? How does one weave the linear, if exponential "great acceleration" into any narrative plot?[18] What about the postmodern claim that "history" does not exist, that there is no grand all-encompassing narrative guaranteeing a sense of history in terms of meaning or direction?[19] If so, not only the end of history is to be announced but that there has never been any history. But even if there is no history, perhaps there can at least be a story or stories?

TELLING THE CHRISTIAN STORY

Christians tell a story of God's economy, that is, the narrative of God's work in the world. This story can be told in rather different ways. It can also be messed up in many different ways. One crucial aspect of this story is that it is open ended; we are telling the story as we are participating in it, with a sense of what has happened in the past, in order to make sense of the present in anticipation of what the future may hold. Christians have no privileged revealed information about how the story will end, if it is even meaningful to speak about such an "end of the world."[20] Whatever they hope for, the future is based not on predictions but on trust in God's loyalty to God's own promises. Such loyalty is based on God's name and on perceptions of God's character, which is distinctive within the Jewish-Christian tradition. Although such hope is oriented toward the future, as a virtue it can only be expressed with regard to the present. The interplay between a mistrusted past, an anxiety-filled present, and an uncertain future is best expressed in the form of narrative.[21]

Christians would want to be alert to what South African theologians typically speak of as a Kairos moment, a moment of truth, a moment when the gospel is at stake, when decisions about life and death, salvation or destruction, orthodoxy or heresy, are at stake.[22] In the context of the Anthropocene it would amount to a failure of nerve not to address the underlying question that those listening to Christian accounts of such a narrative will have: What is God actually doing? What is God up to, given what we know already about climate change? Where can signs of God's presence and creative engagement with the world be found (*vestigia Dei*)? How should we understand divine providence and governance in history (*gubernatio*)? Which countermovements of the Spirit can be discerned?

Views on God's involvement in history have always been extremely dangerous, as is well recognized by the South African theologian Jaap Durand in an essay on "the finger of God" in history, with reference to apartheid theology.[23] Typically, narrow group interests are expressed in answering this question, thus serving as a theological legitimation of such interests. As Susan Neiman observes, "Providence is a tool invented by the rich to lull those whom they oppress into silent endurance. The rich have no need of virtue or faith for their desires are met without them. . . . Providence is either a tool invented for oppression or itself an instrument of injustice."[24] And here is a jibe from Peter Sloterdijk in his history of globalization: "For the Biblicists among the occupiers, America was undoubtedly the ace God had kept up his sleeve for a millennium and a half to play it in the time of greatest need, in the religio-political agony of the Occident. By allowing his Catholic servant Columbus to find America just in time, God used the ploy of divine providence to show his Protestant followers the way in the second exodus."[25] One may retort that such a notion of providence was merely replaced by a trust in the market economy and its production of wealth as an assurance of prosperity, at least for some.[26] Indeed, there is only one thing that is more dangerous than raising the question of God's hand in history and that is failing to raise it at all. Alternative responses remain as distortive as the ones they replace.

This provides little solace though, given the trail of blood and tears that answering this question has left behind.

The question what God is doing is, one would presume, one filled with anxiety. The message of the story, the story of God's economy, is one of salvation addressed to a world clearly in need of salvation. But can this message really make a difference? How is that to be accomplished? Can we even speak about God's actions in the world? How is divine action to be understood?[27] How should the story of what God is doing now be told? Can the story itself still be saved?

To make matters worse, the claims for salvation (the quasi-soteriologies discussed above) have tended to aggravate the underlying problem so that one tends to become suspicious of any such claims. The question is not only how salvation is to come about but also what it is that has to be saved. Offers of "cheap grace" fail to offer the required clarity. "Saving the Earth" may be apt, but this phrase still hides the question whether the focus is on evolving planetary systems, the evolution of species, current species that are at risk of extinction, the (evolution of) human species, some humans more than others, human civilization and its achievements in terms of art, science, and technology, middle-class (consumerist) lifestyles,[28] or even the capitalist world order. What exactly is it that is worth sustaining, if cosmic, biological, and cultural evolution is factored in?[29] The stability of the Holocene? In an ever-changing climate what kinds of change would change everything? The danger of Anthropocene stories is that humanity is not only the main character but marginalizes all other actors who can only react to human impact. With Donna Haraway, all of us need to tell the story in sympoetic ways so that humans are clearly of the Earth and so that the biotic and abiotic powers of the Earth are recognized to be the main story.[30]

APOCALYPTIC AS GENRE FOR THE STORY?

Storytellers need to hold listeners in suspense, but of course they do not really know the answer where the story is heading toward either. Yet, if Christian storytellers literally have no clue, they can hardly keep telling the story and should take cognizance that no one may be listening any longer. They need not be surprised, then, that there are many others telling a different story without reference to the Triune God, with a whole array of new divinities, demons, and monsters. Novelists, film-makers, poets, cartoonists, and astrologists seem perfectly capable of taking over the role of telling the story of where things are going, and do that for entertainment. They do so in rather graphic apocalyptic imagery—the kind that the Hebrew prophets of old were not shy of adopting either, albeit not for entertainment. As John de Gruchy observes, the four horseman of the apocalypse—plague, war, famine, and death—have returned with frightening intensity.[31]

Indeed, there may well be a need to retrieve apocalyptic as a genre, albeit that a coded language is only necessary where surveillance is comprehensive. It cannot be yielded to the prophets of doom and destruction precisely because the biblical genre inspired hope amid oppression and persistent trauma.[32] This is a risky strategy not least because of cheap fear-mongering, but also because future scenarios may easily become self-fulfilling prophecies of the worst possible kind. Many observers note that apocalyptic narratives risk becoming counterproductive. They can lead to denial, apathy, paralysis, fatalism, and the like.[33] Conversely, the doomsday predictions of former apocalyptic prophets, announcing the end of the world, can render us averse when 97 percent of climate scientists issue warnings that provide evidence of changes already happening.[34] Ironically, while secular (e.g., scientific) apocalyptic narratives tend to replace religious narratives as authoritative, religion can stage a disreputable comeback: where statistics seem to only confirm the permanence of destruction, apocalyptic forms of religion offer the hope of postapocalyptic redemption, if only for a few. Indeed, apocalyptic soon becomes very messy indeed.

Where apocalyptic imagery may be helpful is to signal a rupture (not a rapture!) in history—that things cannot just continue as they do. This is the point of recognizing that humans have become a geological force of nature in the Anthropocene. This should stop us in our tracks but need not incite a paralyzing fear. As David Wallace-Wells observes, hope may be more motivating than fear but fear can motivate too: "There is no single way to best tell the story of climate change, no single rhetorical approach likely to work on a given audience and none too dangerous to try."[35] Where the popular use of apocalyptic does become problematic, this is because it leads to forms of escapism, an inability or unwillingness to confront contemporary challenge. In Bruno Latour's imagery this is the challenge of "facing Gaia":

> the fireworks of the Apocalypse are not there to prepare us for an ecstatic elevation towards the Heavens; on the contrary, they are there to keep us from being driven from the Earth as it reacts to our efforts to dominate it. We have misunderstood the injunction: we weren't supposed to bring heaven unto the Earth, but, first, to take care of the Earth, thanks to the Heavens. This is the only way to oblige us to change the direction of our attention after so many years spent neglecting what was taking place behind our backs. . . . To put it baldly: in the face of what is to come, we cannot continue to believe in the old future if we want to have a future at all. This is what I mean by "facing Gaia."[36]

Slavoj Žižek captures the ambiguity of calls for radical change in a formulation worth quoting at some length:

> No wonder, then, that the by far predominant version of ecology is the ecology of fear, fear of a catastrophe—human-made or natural—that may deeply perturb, destroy even, the human civilization, fear that pushes us to plan measures that would protect our safety. This ecology of fear has all the chances of developing into the predominant form of ideology of global capitalism, a new opium for the masses replacing the declining religion: it takes over the old religion's fundamental function, that of putting on an unquestionable authority which can impose limits. The lesson this ecology is constantly hammering is our finitude: we are not Cartesian subjects extracted from reality, we are finite beings embedded in a bio-sphere which vastly transgresses our horizon. In our exploitation of natural resources, we are borrowing from the future, so one should treat our Earth with respect, as something ultimately Sacred, something that should not be unveiled totally, that should and will forever remain a Mystery, a power we should trust, not dominate. While we cannot gain full mastery over our bio-sphere, it is unfortunately in our power to derail it, to disturb its balance so that it will run amok, swiping us away in the process. This is why, although ecologists are all the time demanding that we change radically our way of life, underlying this demand is its opposite, a deep distrust of change, of development, of progress: every radical change can have the unintended consequence of triggering a catastrophe.[37]

Apocalyptic as a genre is therefore not always opportune. At best, it serves the purposes of consolation and fostering resilience to the victims of history when confronted with the overwhelming power of empire. The purpose cannot be more calls to action and certainly not entertainment. There may indeed come a moment in time when prophecy is no longer appropriate, where more calls to act *now*,[38] "before it is too late," would not do because atmospheric trends have become irreversible. Let me illustrate this with one such call to recognize the moment of truth, namely, one issued by Naomi Klein:

> We don't need a separate climate movement; we need to seize the climate moment. The nature of the moment is familiar but bears repeating: whether or not industrialized countries begin deeply cutting our emissions this decade will determine whether we can expect the same from rapidly developing nations like China and India next decade. That, in turn, will determine whether or not humanity can stay within a collective carbon budget that will give us a decent chance of keeping warming below levels that our own governments have agreed are unacceptably dangerous.[39]

It is not that there is anything wrong with such calls. The reality, though, is that this call was written around 2013 and that seven years later such significant emission cuts are not yet evident. Yet, similar calls are issued every year to "seize the moment," to lessen the impact of a changing climate. The year 2020 (when this text was finalized) was declared by Johan Rockström to be a "super year" for international policy on the environment. It is the year when carbon emissions must peak if the target of keeping the rise in global surface temperature below 2 degrees Celsius above preindustrial levels is not to be jeopardized. He urges, "We must make the most of a super-year in 2020 for saving the global commons."[40] Such calls are often accompanied with appropriate advice as to what is indeed feasible, but such calls are typically resisted—also by those who recognize the grave dangers—for all the reasons explored in this diagnosis of what is wrong with the world. What do prophets do, when they realize that their calls will not be heeded? This seems to evoke religious and not merely secular discernment. More often than not it evokes pessimism, nihilism, and resignation. Journalist and essayist Roy Scranton captures the paradoxical situation within which such ethical calls put us:

> We must act *now*, as flawed, failed, flailing selves. At the same time, the situation we find ourselves in is beyond our power to change. The planet *will* get warmer. The ice caps *will* melt. The seas *will* rise. The global fossil-fueled, consumer capitalist civilization we will live in *will* come to an end.[41]

He concludes, "There is no utopia, no Planet B, no salvation, no escape."[42] And then adds this:

> Anyone who pays much attention to politics can assume that we're almost certainly going to botch this challenge. In order to stop emitting waste carbon dioxide completely within the next five or ten years, we would need to radically reorient all human economic and social production, a task that is scarcely imaginable, much less feasible. It would demand centralized control of key economic sectors, massive state investment in carbon capture and sequestration, and global coordination on a scale never before seen, at the very moment when political and economic structures that held the capitalist world order together under American leadership after World War II are splintering, and extremist libertarians are dismantling the United States government from the inside. The very idea of unified national political action toward a single goal seems farcical, and unified action on a global scale mere whimsy. What's more, significant and dangerous levels of warming are already baked in to the system from all the carbon dioxide we've already dumped. There is a time lag between CO_2 increase and subsequent effects, between the wind we sow

> and the whirlwind we reap. Our lives are lived in that gap. My daughter was born there.[43]

For Clive Hamilton, with a bitter sense of irony, hope, the hope that everything will get better, that the worst will be averted, may well be the enemy of action. Indeed, "hoping that a major disruption of the Earth's climate can be avoided is a delusion. Optimism sustained against the facts, including unfounded beliefs in the power of consumer action or in technological rescue, risks turning hope into fantasies."[44] The problem is then that one tends to be suspended between senseless hope and equally senseless despair.[45] Another version comes from Greta Thunberg, the Swedish then sixteen-year-old, speaking truth to power at the annual meeting of the World Economic Forum in Davos in 2019. She dismisses attempts to paint a hopeful scenario:

> But I don't want your hope. I want you to panic. I want you to feel the fear I do. Every day. And I want you to act. I want you to behave like our house is on fire. Because it is.[46]

And at a UN climate action summit in New York on September 23, 2019:

> You have stolen my dreams and my childhood with your empty words. And yet I'm one of the lucky ones. People are suffering. People are dying. Entire ecosystems are collapsing. We are in the beginning of a mass extinction. And all you can talk about is money and fairy tales of eternal economic growth. How dare you.[47]

The deeper reason why such calls for action may well be doomed for failure is that calling on this or that group of people to do this or that or the other thing will not make the difference because the call cannot by itself inspire the required action. It seems that the ethical motivation to do what is required is lacking. A utilitarian sense of acting in enlightened self-interest has become part of the problem.[48] A Kantian sense of doing one's duty could provide a rational account of what to do but does not offer the motivation for doing it. The virtues of loving nature would always have an only limited appeal. Calls to act responsibly require a response to the nearby face of the other/Other (e.g., role responsibility) to prompt action. It therefore seems that we do not have the requisite ethical resources to draw on to face the Anthropocene.[49] As Clive Hamilton observes, we may have to "accept that the Enlightenment did not banish all darkness and that the lamp of Reason shines too dimly to guide us through the night falling over us."[50] Again, if we know what to do but cannot get ourselves to do so, this points in the direction of a spiritual and not only a moral problem.

DISCERNING THE SIGNS OF THE TIME

How, then, do we tell the story? What genre is appropriate? This depends on the discernment of the signs of the time, to recognize the symptoms in order to diagnose the underlying disease. A sense of realism requires us to extrapolate current trajectories. It is indeed unlikely that sufficient political will can be mustered timeously to make drastic cuts in carbon emissions, despite activist campaigns, many good intentions, ample scientific reports, and regular Conferences of the Parties. Catastrophic climate change is now virtually certain, declared Clive Hamilton in *Requiem for a Species* (2010).[51] It is only the degree of catastrophe that can still be engaged with. The truth is that the endless climate negotiations resemble "a form of high-stakes gambling in which catastrophe is the card that is expected to trump all others."[52] It is the denial of the magnitude and gravity of the situation, more than the biogeophysical trends themselves, that render an "apocalypse" plausible.[53] In the sobering comment from Yves Cochet, "So, the collapse seems inevitable not because scientific knowledge of its approach is uncertain but because the social psychology of human life does not allow the right decisions to be made at the right time. We are reaching the limits of the planet in many ways and at the same time. There are often several ways to solve a local or a circumscribed problem, but facing many global dangers together makes the burden of implementing solutions so high that denial is a natural response. It is such denial that ensures that Apocalypse is near."[54]

It is the failure to act upon the knowledge that is widely available that is startling. In the haunting words of Naomi Oreskes and Erik Conway: "While analysts differ on the details, virtually all agree that the people of Western civilization knew what was happening to them but were unable to stop it. Indeed the most startling aspect of the story is just how much these people knew, yet how little they acted upon what they knew."[55] Worse: it is evident that the self-evident character of the threat will not make us change our ways.[56] It is the bourgeois belief in the regularity of the world, that things will always be more or less what they are now, that has inbred a complacency, a madness, a derangement, one that means that large numbers of people will fail to respond despite dire warnings.[57] The democratic system with its short time frame of a few years seems too cumbersome to allow for the kind of radical transformation of the global economy that is required to prevent its collapse.[58] Some would suggest the need for a benevolent dictatorship, but how a transition toward that (and back) could be managed, and how tyranny can then be avoided, is left to the imagination.

One is inclined to say that, as long as life prevails, it can never be "too late" for those alive to make changes and to believe that such changes may and will affect future outcomes. There is something logically odd about saying it is "too little, too late" because this is always expressed with reference to the future, that is, it is too late to do something or to prevent something. But that future keeps shifting. One example comes from a comment by Rajendra Pachauri, the former head of the IPCC on its fourth assessment report (2007—note the date): "If there is no action before 2012, that's too late. What we do in the next two to three years will determine our future. This is the defining moment."[59] If this is true, the failure at Copenhagen (2009) was disastrous, but the more hopeful tone of the Paris Agreement puts the defining moment further into the future by guiding action before and after 2020 when the agreement will come into force.

At the same time, it is possible to discern when it is indeed too late to achieve particular outcomes or to prevent others. As mentioned above, the image often used in literature on climate change is that of peddling a boat in a river just above a waterfall. At some point, it can become too late to prevent the boat from going down the waterfall.[60] Often such discernment has also come too late. If alarmist positions may be problematic, so can lukewarm timidity when it overemphasizes uncertainties, downplays risks, and defend the prevailing social order against policies that may threaten it.[61]

In the history of the Jewish-Christian tradition, there are especially three cases where such discernment was required. The prophet Jeremiah recognized that is too late to prevent the fall of the city of Jerusalem to the Babylonian empire—which happened in 587 BCE—and placed his hope on a God who will start anew. Jesus of Nazareth predicted that Jerusalem will fall again—which happened a generation later in 70 CE due to Roman conquest—and placed his hope in the coming kingdom of heaven. Augustine of Hippo saw the Roman Empire crumbling and placed his hope in the "City of God." Following such imagery, the question is therefore whether we are living in such times, in a period before the fall of "Jerusalem," whatever that may mean. If so, there is a need to realize that the "walls" of Jerusalem (whatever that may mean but Wall Street does come to mind) will indeed fall[62] and a need to place one's hope not in mere survival but elsewhere, in a vision of a society that may emerge from the ruins of "Jerusalem" after a lifetime of "exile."

The so-called fallist movement among tertiary students in South Africa ("Rhodes must fall!"; "Fees must fall!"; "Zuma must fall!"; etc.) indicates that the *hope* (not the fear!) for the fall of "Empire" may be a vibrant and indeed a militant one. One is reminded of the vivid imagery and the joyous celebrations over the fall of Babylon as portrayed in Revelations 18—albeit that kings, businessmen, and sea captains bemoaned her fall and the wealth that it created.

WHAT, THEN, DOES THE FUTURE HOLD?

Overtly, our common future in the twenty-first century and well beyond that will depend upon two main variables, namely, when (in which year) carbon emissions will peak and how rapidly it will decline afterward. The first variable is crucial and renders a quasi-metaphysical significance to this point in human history (i.e., the "super year" 2020). Admittedly, the Anthropocene signals a shift from predictability to volatility so that any predictions would run against its core features. Let me nevertheless be bold in making a few assumptions of where the world may well be at a biblical seventy years from now (when my daughter will be eighty-seven if she is still alive) by extending a few current trends, not allowing for dramatic interventions (volcanic eruptions, asteroids, or exterminating plagues), that could be interpreted as "acts of God," either to destroy the current global order (as punishment?) or to rescue the world from self-destruction.[63]

If sufficient political will cannot be mustered timeously to make drastic cuts in carbon emissions, the carbon budget for the rest of the century (arguably one trillion tons of CO_{2e}) will be spent long before the end of the century, if not already by 2050. Staying within this budget is only possible if national interests become subsidiary to global well-being, and this seems unlikely.[64] The ideal of keeping global warming "well below" 2 degrees Celsius above preindustrial levels would therefore have to be abandoned as "highly unlikely."[65] If so, we are on course for a 3 or 4 degree world at best,[66] while a 5 or 6 degree rise is also possible given the role of various feedback loops (not factored in by the IPCC reports). Due to the lag effect of carbon cycles no one really knows where this will end up. As secular scientists observe, we face an apocalyptic future. It may well be "game over" for the climate.[67] The problem is not just the direct impact of climate change, but that it leads to the intensification of a range of other conflicts, serving as a threat multiplier.[68]

Within this changing climate some trends may become irreversible: hundreds of millions humans could die from hunger and thirst due to crop failure and rivers no longer flowing; devastating pandemics may spread relentlessly, massive migrations of people may no longer be controlled; many low-lying areas would have to be evacuated due to hurricanes, flash floods, and rising sea levels; stock exchanges and insurance companies may well collapse due to the sensitivity of markets to so many disasters; international trade will become increasingly difficult;[69] large parts of the Amazons could burn down; the mass extinctions of species are already evident; the myth of endless economic growth would need to be abandoned; the planet's human populations may reach a peak around 2050 but could drop significantly after that—causing havoc to health and social services, pastoral care (all the funerals!), education, scientific research, and property prices; electricity provision at the regional level will be disrupted often; the world-wide internet could be practically disbanded, with catastrophic impacts for business;[70] organized crime will be rampant as people start looking after themselves; long-distance travel will become increasingly dangerous; mega-sport events will become few and far between. Finding food and drinking water will then be a main activity of people; while home gardens proliferate, hunger and malnutrition will spread. There may remain pockets of affluence with high levels of technology but this will require militarized protection. The continued source of such wealth would be dubious if property prices fall, labor becomes scarce, distribution networks disintegrate, and markets collapse. Communication will still be possible but maintaining high-tech infrastructure across vast distances would be tricky. In short, there is an emerging recognition that, as Roy Scranton observes, "if *Homo sapiens* survives the next millennium, it will be survival in a world unrecognizably different from the one we have known for the last 200,000 years."[71] After a few centuries of belief in human progress, we are now staring at the prospect of a century or more of regress.[72]

Such a scenario requires theological interpretation. Again: What is God actually doing? How should the story of what God is doing be told? What lies beyond the Anthropocene?[73] Should one follow Dante's *Divine Comedy* and abandon hope? For Christians this is only an option if we begin to believe that God has indeed abandoned us (and the covenant with Noah), is making other ominous long-term plans, or if we concur that the divine parent has actually passed away, perhaps for our sake,[74] that our Father is dead, that is, "in heaven" only. Indeed, is God doing anything at all? Is the absence of evidence not evidence of absence? Or should one, instead of looking for a radically transcendent purposeful God, hope for an emergent yet all-powerful "Thermostat" (Gaia) to self-regulate global temperatures?[75] But even if "God" is nothing more than a human construct, an idea (whether good or bad or ugly), a vision for the future, a way of talking about human agendas, then this question still has to be addressed.

The prophetic urge is not to predict the end times, but to speak truth to power in present, penultimate times, to recover the liberating, redeeming message of truth, reconciliation, justice, and peace.[76] Note that the question what God is doing now cannot be postponed to the future as if the Anthropocene is only about the future and not also about the recent past and most significantly the elusive present.[77] Only in hindsight may we realize that the real catastrophe is not in what lies ahead, but in the failures of the present.[78] To read the signs of the time is to recognize that (some of) the carbon that we have released into the atmosphere within the space of a century or two will persist there for thousands of years, altering the climate of the Earth perhaps for hundreds of thousands of years.[79] Eschatology is not (only) about the future either, but about the content and significance of hope—facing up to the present.[80] But what can we hope for if we know now already that there is little promise that the story will have a happy ending? How can a narrative of likely failure mobilize people to do what can still be done?

WHAT IS GOD UP TO?

Instead of offering any affirmative answers, here are a few conflicting and rather blatant possibilities for discussion in response to the question of what God is doing now, in anticipation of what the future may hold:[81]

- God is waiting upon humans to change their hearts, habits, and minds through massive religious awakenings. Like in the days of Noah and after the Babylonian exile, there will be a severe crisis, but God promises to save a small remnant through a miracle that will revert global warming, for example, a series of volcanic eruptions, changes in the ocean currents, cooling off Europe, perhaps introducing a new ice age. Thus, God will create a new humanity, one that has acquired a revised set of virtues.
- No mass conversion and no miracles are to be expected. Instead, by not intervening in the present God is punishing the world's human population for their folly through a deadly plague that will wipe out most of the population. This is rightly interpreted as "God's wrath on extractive civilization, and particularly on its secular salvation narrative of human progress through mastery of nature."[82] This will allow God to start again with a small remnant in the twenty-second century. It may be expected that this will follow the pattern of the oasis of Eden and being expelled from that into the desert, the destruction and desolation of Jerusalem, the exile and the return from exile, and the fall of the Roman Empire and the Renaissance of new forms of civilization.[83] More pertinently, it will follow the dialectic of cross and resurrection, where all the followers of Jesus of Nazareth abandoned him but learned to follow the way of the cross, inspired by the Spirit at Pentecost.[84]
- God is asking Christians to do their best to help address climate change through mitigation and adaption efforts, alongside others, and especially to show solidarity with those struggling to cope with the adverse effects of climate change.[85] We require both sustainability and adaptability. We should not seek any higher purpose but live our lives before God in the world as long as that may last—*etsi Deus non daretur*.
- This is way too timid. As secular commentators agree, only mass social movements can save us now from self-destruction.[86] The current global economic system, based on profiteering through mass production and mass consumption, is self-destructive. Only religion can provide the

countervailing moral vision, moral energy, and moral leadership to inspire people to resist industrialized capitalism.[87] Indeed, a green conversion within Christianity is crucial, precisely given its historical impact. This is what is the Triune God is up to.

- God is asking humans to exercise responsible stewardship of God's good creation.[88] We humans have been entrusted a habitable planet and are the heirs of the rich heritage bestowed on us through the Jewish-Christian-Muslim tradition. We are now called to safeguard human civilization (in every of its major forms) and should utilize the technologies available to adjust the Earth's thermostat through geoengineering. We need to attend to our God-given duty to rule over the Earth. This is God's command, for us, today. If we attend to this collectively, God will send a Messiah (an anointed one) to provide humanity with the miracle (technological) solution they so desperately need.[89]
- It is now clear, more than ever before, that God is not really the God of the poor, oppressed and marginalized. It is not the meek who will inherit the Earth. God is favoring the brave and the clever and will help them to reengineer themselves in order to escape from a doomed Earth through colonizing other habitable planets. After all, it will be awfully sad to see millions of years of evolution wasted if something destroys the Earth. Is this not what divine rapture of the elect few implies?
- God is clearly not really concerned about the world after all. What is ultimately important is the salvation of the soul, becoming one with Absolute Spirit more than discerning the movement of the Spirit through matter, gaining Enlightenment more than understanding the physical properties of light. Gnostic forms of Christianity understood God better than those who focus on what is material, bodily, and earthly—that are subject to decay and futility anyway. The Anthropocene confirms that God is abandoning the Earth and invites those truly enlightened to escape from this earthly vale of tears. The first step may be to move from carbon-based to silicon-based intelligence.[90]
- God has decided to abandon the covenant with Noah and will allow the human species to become extinct,[91] but like in the past God will start again to renew God's creation with previously despised species such as rats and cockroaches taking the lead.[92] This is after all just punishment for irresponsible human destruction.[93] The human game has already begun to play itself out.[94]

It may be noted that God-language is not required to raise these questions or to express these possibilities. God-language does have some sobering effect, though, as it probably rules out some worse scenarios (military subjugation of peoples, global euthanasia, biological warfare to reduce the human population). These questions and options arise no matter how secular one's orientation may be.

For Christian theology, the question then remains: Can this story still be held together as one narrative or are there not only different versions of the story but also different stories and indeed different gods? Or is God the Grand Recounter, the storyteller who weaves us as characters into a singular storyline?[95] For Christians the different versions of the story, the different perspectives from which it may be told, even the nineteenth-century death of monotheism, need not be of any concern. Christians do not believe in God for the sake of believing in any available God. It is not as if one first needs to provide an account of the plausibility of God-talk and the very notion of a divine being before such questions can be addressed, as if it is possible to outline a generic notion of "God," if this is indeed the best English term to use for something like the ultimate mystery of the world.[96] The theological question is which of these options is in line with the character of the Triune God in whom Christians say they put their trust.

From the biblical roots of the Christian tradition, it seems, to me at least, unlikely that some form of dramatic divine intervention is to be expected that would miraculously avert a climate catastrophe. Despite the vehement cry from the crucified Christ that his Father has abandoned him, the Father did not intervene. This nonintervention made room for the slow-moving advocacy of God's Spirit to transform all things. It is unlikely that this is the message a world in need of salvation from self-destruction would wish to hear. Is this gospel then good news? This depends upon the way the story is told by Christian witnesses.[97] The authenticity of this confession will be tested in decades to come amid a changing climate.

NOTES

1. See Thomas R. Dunlap, "Environmentalism, a Secular Faith," *Environmental Values* 15:3 (2006): 321–30.
2. It is striking how often books on the Anthropocene recognize the role of such cosmogenic narratives. One example may suffice, namely, Ellis's *Anthropocene: A Very Short Introduction* (2018) where the role of such stories is highlighted in the opening pages.
3. Serres, *The Incandescent*, 195.
4. Sloterdijk, *In the World Interior of Capital*, 176.
5. See Lewis and Maslin, *The Human Planet*, 399.
6. See the fine essay by Clingerman, "Time's Arrow and Narratives of Climate Engineering." He notes the ability of narratives to provide a framing structure by gathering disparate events, facts, and values into a meaningful temporal sequence (90). Yet stories are also partial, fragmentary, and limited in perspective (93).

Clingerman's concern is the tacit assumptions of the master narrative assumed by geoengineering. The theological question is indeed not whether we tell the story of the Anthropocene, but how that story is told and how God (God's identity and character) is portrayed as a character in that story.

7. See Hamilton, Bonneuil and Gemene (eds), *The Anthropocene and the Global Environmental Crisis*, 5.

8. Timothy Morton offers a decolonial critique of an anthropocentric notion of space and place: "Place now has nothing to do with good old reliable constancy. What has dissolved is the idea of constant presence: the myth that something is real insofar as it is consistently, constantly 'there.' The concept space was always a constant-presencing machine for making things appear consistent and solid, to make them easier to colonize, enslave, and plunder. Constant presence was part of the anthropocentric colonization protocol." See Morton, *Dark Ecology*, 10.

9. Latour, *Facing Gaia*, 135.

10. Ibid., 73.

11. This may not satisfy Latour or Peter Sloterdijk on whom he draws. Sloterdijk offers a critique of the old obsession with the form of the Globe, as if we can adopt a holistic perspective on the globe. For him, we need to pass from monotheism to monogeism, that is, those have no spare planet and thus have only one Earth but "do not know its form any better than they know the face of their former God." See Latour, *Facing Gaia*, 145.

12. Holmes Rolston (III), *Three Big Bangs: Matter-Energy, Life, Mind* (New York: Columbia University Press, 2010).

13. Serres, *The Incandescent*, 103.

14. See Bonneuil and Fressoz, *The Shock of the Anthropocene*, 47–64.

15. See Bonneuil, "The Geological Turn," 17–18.

16. Ibid., 27.

17. See Ibid., 29. In the discussion below I will follow Bonneuil's third, eco-catastrophic narrative, although the discernment of God's "finger" in the Anthropocene covers the fuller spectrum of Bonneuil's narratives.

18. For this question, see Ten Bos, *Dwalen in het Antropoceen*, 124, following insights from Maurice Blanchot. Ten Bos likens die Anthropocene to a catastrophe that is by definition unprecedented. Instead, one may also see it as an underlying tendency of human civilizations since the agricultural and industrial revolutions that became manifest only recently.

19. See Žižek, *Living in the End Times*, 184.

20. See Timothy Morton's discussion of "the end of the world." He argues that "the world" has already ended as a result of the appearance of hyperobjects. This is also a devastating blow against any notion of teleology. He says: "The end of teleology is the end of the world. This end is precisely not an instant vaporization, but rather a lingering coexistence of strange strangers. For the end of the world is the end of endings, the end of telos, and the beginning of an uncertain, hesitating futurality." See Morton, *Hyperobjects*, 95.

21. One of the best expositions of this interplay remains the three volumes of Paul Ricoeur's *Time and Narrative* (Chicago: University of Chicago Press, 1984, 1985, 1988).

22. The need to distinguish between the Anthropocene as *Chronos* and as *Kairos* is recognized by Michael Northcott, "Eschatology in the Anthropocene," in *The Anthropocene and the Global Environmental Crisis*, edited by Clive Hamilton, Christophe Bonneuil and François Gemene (London and New York: Routledge, 2015), 100–11.

23. See again Durand, "God in History: An Unresolved Problem."

24. Neiman, *Evil in Modern Thought*, 182. Also quoted in Hamilton, "The Theodicy of the 'Good Anthropocene,'" 236.

25. Sloterdijk, *In the World Interior of Capital*, 117.

26. Van Hoogstraaten, *Deep Economy*, 127–28.

27. See, especially, Robert John Russell, Nancey Murphy and William R. Stoeger (eds), *Scientific Perspectives on Divine Action: Twenty Years of Challenge and Progress* (Berkeley: Center for Theology and the Natural Sciences / Vatican City: Vatican Observatory Publications, 2008). I have offered a survey of discourse on divine action in Afrikaans in *Lewend en Kragtig?* (Wellington: Bybelmedia, 2010). See also Chapter 4 of *The Earth in God's Economy: Creation, Salvation and Consummation in Ecological Perspective* (Berlin: LIT Verlag, 2015).

28. Naomi Klein's comment is apt in this regard: "As for pitching climate action as a way to protect America's high-consumerist 'way of life'—that is either dishonest or delusional because a way of life based on the promise of infinite growth cannot be protected, least of all exported to every corner of the globe." See Klein, *This Changes Everything*, 58.

29. I have argued above that this is the crucial question for discourse on sustainability: What is it that is worth sustaining, and not so much the questions whether, how, or how long that can be sustained. See again my "Is it not God's Mercy That Nourishes and Sustains Us . . . Forever?"

30. See Haraway, "Staying with the Trouble," 59.

31. See De Gruchy, *The End is Not Yet*, 10.

32. Following a comment by Slavoj Žižek that there are contexts where trauma is persistent, where it is not the specter of trauma that remains but the trauma itself, so that speaking of posttraumatic stress is meaningless. See *Living in the End Times*, 293.

33. See the argument of Giovanni Bettini in "(In)convenient Convergences – 'Climate Refugees,' Apocalyptic Discourses and the Depoliticization of the Debate on Climate-induced Migration," in *(De)constructing the Greenhouse: Interpretive Approaches to Global Climate Governance*, edited by Chris Methmann, Rolf Delf Rothe and Benjamin Stephan (London: Routledge, 2013), 122–36.

34. See Wood, *Deep Time, Dark Times*, 55.

35. Wallace-Wells, *The Uninhabitable Earth*, 157.

36. Latour, *Facing Gaia*, 244–45. The expression "thanks to the Heavens" is somewhat mischievous: does it refer to an enabling atmosphere, to transcendence or to (the death of) God?

37. See Slavoj Žižek, "Censorship Today: Violence, or Ecology, as New Opium for the Masses," https://www.lacan.com/zizecology1.htm (accessed December 10, 2019).

38. Timothy Morton finds an overbearing metaphysics of presence not only in timekeeping devices but also in any call to act *now*. To him this sounds pretty much like saying: "The point is to stop thinking, stop reflecting, go out and act!" See Morton, *Hyperobjects*, 92.

39. Klein, *This Changes Everything*, 153.

40. Johann Rockström, "Three Steps to Meeting the Climate and Nature Emergency," *The Telegraph*, July 16, 2019, https://www.telegraph.co.uk/business/how-to-be-green/climate-and-nature-emergency/ (accessed May 16, 2020).

41. Scranton, *We're Doomed. Now What?* 68 (his emphasis).

42. Ibid., 325.

43. Ibid., 320–21.

44. Hamilton, *Requiem for a Species*, 132.

45. See Latour, *Facing Gaia*, 242.

46. See https://www.weforum.org/agenda/2019/01/our-house-is-on-fire-16-year-old-greta-thunberg-speaks-truth-to-power?utm_source=newsletter&utm_medium=email&utm_content=hier&utm_campaign=VrydagNuus%202019 (accessed February 7, 2019).

47. See https://www.climatechangenews.com/2019/09/23/dare-greta-thunbergs-tearful-speech-leaders-un-video/ (accessed September 24, 2019).

48. For a discussion, see Gardiner, *A Perfect Moral Storm*, 235–45.

49. Hamilton, *Defiant Earth*, 155.

50. Ibid., 156.

51. Hamilton, *Requiem for a Species*, xiv.

52. Ghosh, *The Great Derangement*, 149.

53. See Yves Cochet, "Green Eschatology," in *The Anthropocene and the Global Environmental Crisis*, edited by Clive Hamilton, Christophe Bonneuil and François Gemene (London and New York: Routledge, 2015), 112–19 (113).

54. Cochet, "Green Eschatology," 118–19.

55. See Naomi Oreskes and Erik Conway, *The Collapse of Western Civilization: A View from the Future* (New York: Columbia University Press, 2014), 1–2.

56. As noted by Latour, *Facing Gaia*, 73.

57. This is the basic thesis of novelist Ghosh in *The Great Derangement*, 36.

58. Cochet, "Green Eschatology," 115.

59. Quoted in Gardiner, *A Perfect Moral Storm*, 81.

60. Steffen et al., *Global Change and the Earth System*, 294.

61. See Hamilton, *Earthmasters*, 112.

62. This was my argument in an unpublished presentation provocatively entitled, "Why I am telling my children that they should expect the 'walls' of 'Jerusalem' to fall within their envisaged lifetime," Winter school, Stellenbosch University, July 16, 2019. Like Johan Rockström, I suggested that 2020 may be the year that determines my children's future. I did not see Covid-19 coming!

63. See the interesting exercise by Naomi Oreskes and Erik M. Conway to imagine how a future historian living in the Second People's Republic of China in 2393 may look back at a past that is our present and (possible) future in *The Collapse of Western Civilization*. They anticipate 2093 as the year of the "Great Collapse," that is, when

the West Antarctica Ice Sheet will have collapsed, driving up sea levels by 5 meters across the globe.

64. See the comment by Michel Serres: "We will be saved from the apocalypse if and only if all the humans of all the countries unite without borders to make the world their partner." See Serres, *Malfeasance*, 84.

65. This formulation hints at the language of the Paris Agreement of 2015 which seeks to limit global warming to "well below" 2 degree Celsius and to "pursue efforts" of limiting global warming to 1.5 degree Celsius. In the language of the United Nations Framework Convention on Climate Change, "more likely than not" implies a greater than 50 percent chance, "likely" a greater than 66 percent chance, and "very likely" a greater than 90 percent chance.

66. Kevin Anderson from the Tyndall Centre for Climate Change Research states bluntly that global warming of 4 degrees Celsius is "incompatible with any reasonable characterization of an organized, equitable and civilized global community." Quoted in Klein, *This Changes Everything*, 13.

67. See James Hansen, "Game Over for the Climate," https://www.nytimes.com/2012/05/10/opinion/game-over-for-the-climate.html (accessed September 13, 2018).

68. See Ghosh, *The Great Derangement*, 143.

69. This is how Mark Lynas describes changes to economic markets in a "five degrees world": "For humanity, a new era of enforced localism is likely, where globalisation goes into reverse and people reassert more restricted identities. Our economy is globally interconnected at present, with huge volumes of trade taking place between far-flung regions. But hypothetical customers in some ravaged coastal city of the future will no longer be able to buy, whilst producers in a drought-stricken subtropical zone will no longer have anything to sell. Well before this situation is reached, the sensitive and volatile capital markets will surely have collapsed, erasing ownership bods between foreign and domestic capital, and precipitating a world-wide economic depression." See Lynas, *Six Degrees*, 210.

70. See Cochet, "Green Eschatology," 114.

71. Scranton, *Learning to Die in the Anthropocene*, 19.

72. Hamilton, *Requiem for a Species*, 209.

73. See Bergmann's similar questions: "Is there space to imagine a new geological era beyond the Anthropocene? Maybe the 'Eocene,' where human and other life forms cohabitate at Earth in fully just and peaceful entanglements? Or rather an era of apocalypse where humans eradicate themselves from the planet, followed by an era of new genesis where evolution searches for new paths without human intervention? Or a 'post-technocene' where the fetishization of money and machines has been overcome and technical spaces have turned into lived spaces?" See Sigurd Bergmann, "Theology in the Anthropocene – and beyond?" in *Contextual Theology: Skills and Practices of Liberating Faith*, edited by Sigurd Bergmann and Mika Vähäkangas (London et al.: Routledge, 2020, forthcoming).

74. This is the kind of God Slavoj Žižek prefers for the radical left: "A God who has fully 'become a man,' a comrade who not only 'does not exist' but also *knows this himself*, accepts his own erasure, passing over entirely into the love that binds all members of the 'Holy Ghost,' that is, of the Party or emancipatory collective." See the final few lines of *Living in the End Times*, 402.

75. See Latour, *Facing Gaia*, 132.

76. De Gruchy, *The End is Not Yet*, 148.

77. See Ghosh, *The Great Derangement*, 72.

78. See Scranton, *We're Doomed. Now What?* 329.

79. See Hamilton, *Earthmasters*, 193. Drawing on the work of paleoclimatologists, Hamilton notes that releasing an additional 500,000 tons of carbon into the atmosphere will suppress the next ice age (due in 50,000 years), while burning all the known fossil fuel reserves will also suppress the one after that (due in 130,000 years), altering the pattern for the next half a million years. He concludes: "Our activities have so changed the climatic future that we have prevented one and possibly several ice ages. I think it will take decades for us to understand the enormous implications and meaning of this fact. The Earth will take tens or hundreds of thousands of years to reach a new equilibrium following the pulse of carbon emissions sent into the atmosphere by humans mostly over the century from 1950 to 2050. Only then could the era of human-induced global warming come to an end" (193).

80. See also Latour, *Facing Gaia*, 219. I have explored this at some length in my Ernst M. Conradie, *Hope for Earth: Vistas on a New Century* (Eugene: Wipf & Stck, 2005).

81. A discussion of these options is also found in my essay Ernst M. Conradie, "The Story of God's Work: An Open-ended Narrative," in *T&T Clark Handbook on Christian Theology and Climate Change*, edited by Ernst M. Conradie and Hilda Koster (London: T&T Clark, 2019), 462–66. The two texts were written simultaneously.

82. See Jenkins, Berry and Kreider, "Religion and Climate Change," 9.2, with reference to Michael Northcott's position as adopted in *A Moral Climate*.

83. Augustine realized that the fall of the empire will undermine social stability but that it will not be protected by God's providence. He also knew that "the reason for the collapse of the empire was the greed, corruption, and arrogance of Rome, its endless wars, political infighting, and the failure of justice. The 'wrath of God,' as the Hebrew prophets knew, was a metaphor for the inevitable outcome of such human disobedience." See De Gruchy, *The End is Not Yet*, 49. De Gruchy adds that this serves as a warning that empires fall not because they are attacked by enemies but because of their own stupidity and folly, their moral failure, and the lack of moral leadership (50).

84. Clive Hamilton offers a secular version of such a narrative. He asks whether humankind can be redeemed. Will we be given a chance for a second civilization? Will a "resurrected" (his term) humanity emerge from the planetary ashes of the old civilization? He notes that there is no guarantee for this eventuality or that he lessons will be learned from the past, even if they were to declare "Never again!" See the very last words of Hamilton's, *Defiant Earth*, 162.

85. See the booklet produced by the World Council of Churches, *Solidarity with Victims of Climate Change* (Geneva: WCC, 2002). Its haunting argument is that such solidarity will be required in a time "when love grows cold."

86. See the final conclusion in Naomi Klein, *This Changes Everything*, 450.

87. See again Tucker and Grim, "Foreword," xix.

88. See the critique of such "stewardship of Empire" by Yusoff, *A Billion Black Anthropocenes or None*, 27: "The colonial assumption for the responsibility for and

of the world is articulated anew as the white man's burden—a paternalism that is tied to a redemptive narrative of saving the world from harm on account of others while maintaining the protective thick skin of innocence."

89. For critical commentary on this option, see the chapter by Naomi Klein on billionaire messiahs, in *This Changes Everything*, 230–55.

90. Bruno Latour retorts: "What use is it if you save your soul, if it means losing the world?" See Latour, *Facing Gaia*, 211.

91. In the words of Richard Leakey: "Homo sapiens might not only be the agent of the sixth extinction, but also risks being one of its victims." Quoted in Kolbert, *The Sixth Extinction*, 268. The reference is to Richard E. Leakey and Roger Lewin, *The Sixth Extinction: Patterns of Life and the Future of Humankind* (New York: Anchor, 1996), 249. Yet, visions of human extinction may be a form of escapism as we then need not take any further responsibility for the future.

92. Elizabeth Kolbert closes her book *The Sixth Extinction* with these words: "Extinction will continue to determine the course of life long after everything people have written and painted and built has been ground into dust and giant rats have—or have not inherited the earth" (269).

93. See Peter Scott, "God's Work Through the Emergence of Humanity," in *T&T Clark Handbook on Christian Theology and Climate Change*, edited by Ernst M. Conradie and Hilda Koster (London: T&T Clark, 2019), 373–83: "A final decentring emerges: the end of the human as a species through an extinction event to which the human in a specific society has contributed. We are invited to consider a world without the human. This takes us, I think, to the heart of Christian anthropocentrism. A theological response to such an invitation would need to consider God's purposes in creating and the destiny of creation" (383). Clive Hamilton observes that such a fatalistic acceptance of human extinction is more resonant with an Old Testament pre-occupation with judgment than with a New Testament compassion with the fallenness of humankind. See Hamilton, *Defiant Earth*, 115.

94. See the subtitle of Bill McKibben's recent book *Falter* (2019).

95. See Serres, *The Incandescent*, 20.

96. The two terms that are most widely used in English to translate יהוה (Yahweh), אֱלֹהִים (Elohim), and θεός (Theos) in the biblical roots of Christianity are "Lord" (no comment!) and "God." The etymological roots of the latter are contested but may perhaps be traced to the proto-Germanic guthan, which is also the source of the Dutch god, the German Gott, the Old Norse guð, and the Gothic guþ). The Online Etymology Dictionary suggests that ghutan may be derived from the Proto-Indo-European "ghut-," meaning "that which is invoked"—a sense of awe. See https://www.etymonline.com/word/god (accessed January 20, 2019).

97. In years to come I hope to address this question more fully through a series of twelve edited volumes entitled *An Earthed Faith: Telling the Story amid the Anthropocene*.

Bibliography

Ackoff, Russell L. *Redesigning the Future*. London: Wiley, 1974.

Angus, Ian. *Facing the Anthropocene: Fossil Capitalism and the Crisis of the Earth System*. New York: Monthly Review Press, 2016.

Antonaccio, Maria. "De-moralizing and Re-moralizing the Anthropocene." In *Religion in the Anthropocene*, edited by Celia E. Deane-Drummond, Sigurd Bergmann, and Markus Vogt, 121–37. Eugene: Cascade Books, 2017.

Barth, Karl. *Church Dogmatics IV.1*. Edinburgh: T&T Clark, 1956.

Bartholomew (Ecumenical Patriarch). "Statement for the WCC Working Group on Climate Change." In *Climate Change*. Geneva: WCC, 2005.

Bavinck, Herman. *Gereformeerde Ethiek,* edited by Dirk van Keulen. Utrecht: KokBoekencentrum, 2019.

Béguin-Austin, Midge (ed.). *Sustainable Growth – Contradiction in Terms?* Geneva: The Ecumenical Institute, 1993.

Bergmann, Sigurd. "Theology in the Anthropocene – and Beyond?" In *Contextual Theology: Skills and Practices of Liberating Faith,* edited by Sigurd Bergmann and Mika Vähäkangas. London et al.: Routledge, 2020, forthcoming.

Bettini, Giovanni. "(In)convenient Convergences – 'Climate Refugees', Apocalyptic Discourses and the Depoliticization of the Debate on Climate-induced Migration." In *(De)constructing the Greenhouse: Interpretive Approaches to Global Climate Governance*, edited by Chris Methmann, Rolf Delf Rothe and Benjamin Stephan, 122–36. London: Routledge, 2013.

Biermann, Frank et al. "Down to Earth: Contextualizing the Anthropocene." *Global Environmental Change* 39 (2016): 341–50.

Biko, Steven Bantu. *I Write what I Like: Selected Writings*. Johannesburg: Picardo Africa, 2017.

Boesak, Allan A. *Afscheid van de Onschuld: Een Sociaal-etische Studie over Zwarte Theologie en Zwarte Macht*. Kampen: JH Kok, 1977.

———. "He Made us All, but …" In *Apartheid is a Heresy*, edited by John W. de Gruchy and Charles Villa-Vicencio, 1–9. Cape Town: David Philip, 1983.

———. *Pharaohs on Both Sides of the Blood-Red Waters*. Eugene: Cascade Books, 2017.

Boff, Leonardo. *Cry of the Earth, Cry of the Poor*. Maryknoll: Orbis, 1997.

Bonhoeffer, Dietrich. *Creation and Fall: A Theological Exposition of Genesis 1–3,* edited by John W. de Gruchy. Minneapolis: Fortress Press, 1997.

Bonneuil, Christophe. "The Geological Turn: Narratives of the Anthropocene." In *The Anthropocene and the Global Environmental Crisis*, edited by Clive Hamilton, Christophe Bonneuil and François Gemene, 17–31. London and New York: Routledge, 2015.

Bonneuil, Christophe and Jean-Baptiste Fressoz. *The Shock of the Anthropocene: The Earth, History and Us*. London: Verso, 2017.

Bosch, David J. "Nothing but a Heresy". In *Apartheid is a Heresy*, edited by John W. de Gruchy and Charles Villa-Vicencio, 24–37. Cape Town: David Philip, 1983.

———. "The Roots and Fruits of Afrikaner Civil Religion." In *New Faces of Africa: Essays in Honour of Ben (Barend Jacobus) Marais*, edited by Jan W. Hofmeyr and Willem S. Nicol, 14–35. Pretoria: Unisa, 1984.

Botman, H. Russel. "Narrative Challenges in a Situation of Transition." In *To Remember and to Heal: Theological and Psychological Reflections on Truth and Reconciliation,* edited by H. Russel Botman and Robin M. Petersen, 37–44. Cape Town: Human and Rousseau, 1996.

Bouma-Prediger, Steven. *Earthkeeping and Character: Exploring a Christian Ecological Virtue Ethic*. Grand Rapids: Baker, 2019.

Callicot, J. Baird. *Thinking Like a Planet: The Land Ethics and the Earth Ethic*. Oxford: Oxford University Press, 2013.

Calvin, John. *Institutes of the Christian Religion*, edited by John T. McNeill, translated by Ford Lewis Battles. Louisville: Westminster John Knox Press, 1960.

Camara, Dom Helder. *Spiral of Violence*. London: Sheed and Ward, 1971.

Chakrabarty, Dipesh. "Postcolonial Studies and the Challenge of Climate Change." *New Literary History* 43:1 (2012): 1–18.

———. "The Anthropocene and the Convergence of Histories." In *The Anthropocene and the Global Environmental Crisis*, edited by Clive Hamilton, Christophe Bonneuil, and François Gemene, 44–56. London and New York: Routledge, 2015.

———. "The Climate of History: Four Theses." *Critical Inquiry* 35:2 (2009): 197–222.

———. "The Politics of Climate Change is More than the Politics of Capitalism." *Theory, Culture & Society* 34:2–3 (2017): 25–37.

Church of Norway. *The Consumer Society as an Ethical Challenge*. Oslo: Church of Norway, 1995.

Churchman, C. West. "Wicked Problems." *Management Science* 14:4 (1967): 141–42.

Clingerman, Forrest and Kevin J. O'Brien (eds). *Theological and Ethical Responses to Climate Engineering*. Lanham: Lexington, 2016.

Clingerman, Forrest and Kevin J. O'Brien. "Playing God: Why Religion Belongs in the Climate Engineering Debate." In *Theological and Ethical Responses to Climate Engineering*, edited by Forrest Clingerman and Kevin J. O'Brien, 3–15. Lanham: Lexington, 2016.

Clingerman, Forrest. "Time's Arrow and Narratives of Climate Engineering." In *Theological and Ethical Responses to Climate Engineering*, edited by Forrest Clingerman and Kevin J. O'Brien, 81–101. Lanham: Lexington, 2016.

Cobb, John B. (jr). "Liberation Theology and the Global Economy." In *Liberating the Future: God, Mammon and Theology*, edited by Joerg Rieger, 27–42. Minneapolis: Fortress Press, 1998.

Cochet, Yves. "Green Eschatology." In *The Anthropocene and the Global Environmental Crisis*, edited by Clive Hamilton, Christophe Bonneuil, and François Gemene, 112–19. London and New York: Routledge, 2015.

Coetzee, Murray H. and Ernst M. Conradie. "Apartheid as Quasi-soteriology: The Remaining Lure and Threat." *Journal of Theology for Southern Africa* 138 (2010): 112–23.

Conradie, Ernst M. (ed.). *Reconciliation as a Guiding Vision for South Africa?* Stellenbosch: SUN Press, 2013.

Conradie, Ernst M. "All Theology is Natural Theology: The Hermeneutic Necessity of Natural Theology?" *Ned Geref Teologiese Tydskrif* 52:1&2 (2011): 58–65.

———. "Climate Justice, Food Security … and God: Some Reflections from the Perspective of Ecotheology." In *Climate Justice and Food Security: Theological Education and Christian Leadership Development*, edited by Dietrich Werner and Elizabeth Jeglitzka, 109–34. Geneva: Globethics.net, 2016.

———. "Confessing Guilt in the Context of Climate Change: Some South African Perspectives." In *Ecological Awareness: Exploring Religion, Ethics and Aesthetics*, edited by Sigurd Bergmann and Heather Eaton, 77–96. Berlin: LIT Verlag, 2011.

———. "Could Eating other Creatures be a Way of Discovering their Intrinsic Value?" *Journal of Theology for Southern Africa* 164 (2019): 26–39.

———. "Divine Election and Migration: The Worst Possible Way to Address the Predicament of Refugees?" *Stellenbosch Theological Journal* 2:1 (2016): 131–48.

———. "Do Only Humans Sin? In Conversation with Frans de Waal." In *Are We Special? Science and Theology Questioning Human Uniqueness*, edited by Michael Fuller and Dirk Evers, 117–35. Heidelberg: Springer, 2017.

———. "Eat and/or be Eaten: The Evolutionary Roots of Violence?" *Scriptura* 114 (2015): 1–22.

———. "From Land Reform to Poo Protesting: Some Theological Reflections on the Ecological Repercussions of Economic Inequality." *Scriptura* 113 (2014): 1–16.

———. "Globalisation, Consumerism and the Call for a *Status Confessionis*." In *Globalisation Volume II: Global Crisis, Global Challenge, Global Faith – An Ongoing Response to the Accra Confession*, edited by Allan A. Boesak and Len Hansen, 53–76. Stellenbosch: SUN Press, 2010.

———. "Is it not God's Mercy That Nourishes and Sustains Us … Forever? Some Theological Perspectives on Entangled Sustainabilities." *Scriptura* 116 (2017): 38–54.

———. "John Haught on Original Sin: A Conversation." *HTS Teologiese Studies / Theological Studies* 72:4 (2016): 1–10.

———. "On Social Evil and Natural Evil: In Conversation with Christopher Southgate." *Zygon: Journal of Religion and Science* 53:3 (2018): 752–65.

———. "Penultimate Perspectives on the Root Causes of Environmental Destruction in Africa." *Scriptura* 115 (2016): 1–19.

———. "The Christian Faith and Evolution: An Evolving, Unresolved Debate." *Verbum et Ecclesia* 39:1 (2018): 1–13.

———. "The Emergence of Human Sin." In *T&T Clark Handbook on Christian Theology and Climate Change*, edited by Ernst M. Conradie and Hilda P. Koster, 384–94. London et al.: T&T Clark, 2019.

———. "The Story of God's Work: An Open-ended Narrative." In *T&T Clark Handbook on Christian Theology and Climate Change*, edited by Ernst M. Conradie and Hilda Koster, 462–66. London: T&T Clark, 2019.

———. "To Cover the Many Sins of Galamsey Mining." *Missionalia* 46:1 (2018): 109–30.

———. "Twelve Theses on the Place of Christian Theology in Multi-disciplinary Conversations." *Stellenbosch Theological Journal* 1:1 (2015): 375–86.

———. "What Diagnosis? Which Remedy? Critical Reflections on the Diagnostic Overview of South Africa's National Planning Commission." *Scriptura* 117 (2018): 1–21.

———. "What Do We Do when We Eat? Part 1: An Inconclusive Inquiry & Part 2: A Theological Inquiry." *Scriptura* 115 (2016): 1–17.

———. "What is the Ecological Significance of God's Transcendence?" In *Nature and Beyond: Transcendence and Immanence in Science and Religion*, edited by Michael Fuller et al. Issues in Science and Theology, 87–99. Heidelberg: Springer, 2020.

———. "What Makes the World Go Round? Some Reformed Perspectives on Pneumatology and Ecology." *Journal of Reformed Theology* 6 (2012): 294–305.

———. "Why Can't the Term Development Just be Dropped Altogether? Some Reflections on the Concept of Maturation as Alternative to Development Discourse." *HTS Theological Studies* 72:4 (2016): 1–11.

———. *Christianity and a Critique of Consumerism: A Survey of Six Points of Entry*. Wellington: Bible Media, 2009.

———. *Hope for Earth: Vistas on a New Century*. Eugene: Wipf & Stock, 2005.

———. *Lewend en Kragtig? In Gesprek oor God se Handelinge in die Wêreld*. Wellington: Bybelmedia, 2010.

———. *Om Reg te Stel: Oor Regverdiging én Geregtigheid*. Wellington: Bybelmedia, 2018.

———. *Redeeming Sin? Social Diagnostics amid Ecological Destruction*. Lanham: Lexington Books, 2017.

———. *Saving the Earth? The Legacy of Reformed Views on "Re-creation"*. Berlin: LIT Verlag, 2013.

———. *The Earth in God's Economy: Creation, Salvation and Consummation in Ecological Perspective*. Berlin: LIT Verlag, 2015.

Conradie, Ernst M., Sigurd Bergmann, Celia E. Deane-Drummond, and Denis Edwards (eds). *Christian Faith and the Earth: Current Paths and Emerging Horizons in Ecotheology*. London: T&T Clark, 2014.

Cox, Harvey. "The Market as God." *The Atlantic Monthly* 283:3 (1999): 18–23.

———. *The Market as God.* Cambridge: Harvard University Press, 2016.

Crist, Eileen. "On the Poverty of our Nomenclature." In *Anthropocene or Capitalocene? Nature, History and the Crisis of Capitalism*, edited by Jason W. Moore, 14–33. Oakland: PM Press, 2016.

Crutzen, Paul J. "Albedo Enhancement by Stratospheric Sulfur Injections: A Contribution to Resolve a Policy Dilemma?" *Climatic Change* 77 (2007): 211–19.

Crutzen, Paul J. and Will Steffen. "How Long Have We Been in the Anthropocene Era?" *Climatic Change* 61 (2003): 251–57.

Dalby, Simon. "Framing the Anthropocene: The Good, the Bad, and the Ugly." *The Anthropocene Review* 3:1 (2016): 33–51.

De Gruchy, John W. "Towards a Confessing Church: The Implications of a Heresy." In *Apartheid is a Heresy*, edited by John W. de Gruchy and Charles Villa-Vicencio, 75–93. Cape Town: David Philip, 1983.

———. *The End is Not Yet: Standing Firm in Apocalyptic Times.* Minneapolis: Fortress, 2017.

De Gruchy, John W. and Charles Villa-Vicencio (eds). *Apartheid is a Heresy.* Cape Town: David Philip, 1983.

De Gruchy, Steve M. *Keeping Body and Soul Together: Reflections by Steve de Gruchy on Theology and Development.* Pietermaritzburg: Cluster Publications, 2015.

De Waal, Frans. "Who Apes Whom." http://www.nytimes.com/2015/09/15/opinion/who-apes-whom.html? (accessed November 6, 2016).

Descola, Philippe. *Beyond Nature and Culture.* Chicago: University of Chicago Press, 2013.

Diamond, Jared. *Collapse: How Societies Choose to Fail or Survive.* New York: Penguin Books, 2005.

Dunlap, Thomas R. "Environmentalism, a Secular Faith." *Environmental Values* 15:3 (2006): 321–30.

Durand, Jaap. "A Confession – Was it Really Necessary?" In *A Moment of Truth: The Confession of the Dutch Reformed Mission Church*, edited by G. Daniel Cloete and Dirk J. Smit, 33–41. Grand Rapids: Eerdmans, 1984.

———. "Die Kerk se Profetiese Roeping." In *Perspektief op die Ope Brief*, edited by David J. Bosch, Adrio König, and Willem Nicol, 66–85. Kaapstad: Human & Rousseau, 1982.

———. "God in History: An Unresolved problem." In *The Meaning of History*, edited by Adrio König and Marie Henri Keane, 171–78. Pretoria: Unisa, 1980.

Edwards, Denis. *Deep Incarnation: God's Redemptive Suffering with Creatures.* Maryknoll: Orbis, 2019.

Ehrlich, Paul R. and John P. Holdren. "Impact of Population Growth." *Science: American Association for the Advancement of Science* 171 (1971): 1212–17.

Ekins, Paul and Michael Jacobs. "Environmental Sustainability and the Growth of GDP: Conditions for Compatibility." In *The North, the South and the Environment: Ecological Constraints and the Global Economy*, edited by Vinit Bhaskar and Andrew Glyn, 9–46. London: Earthscan Publications, 1995.

Ellis, Erle C. "Anthropogenic Transformation of the Terrestrial Biosphere." *Philosophical Transactions of the Royal Society* 369 (2011): 1010–35.

———. "Nature for the People: Toward a Democratic Vision for the Biosphere." *Breakthrough Journal* 7 (2017).

———. "Neither Good Nor Bad." *New York Times*, May 23, 2011. https://www.nytimes.com/roomfordebate/2011/05/19/the-age-of-anthropocene-should-we-worry/neither-good-nor-bad (accessed April 12, 2019).

———. "The Planet of No Return: Human Resilience on an Artificial Earth." *Breakthrough Journal* 2 (2011): 39–44.

———. *Anthropocene: A Very Short Introduction.* Oxford: Oxford University Press, 2018.

Ellis, Erle C. et al. "Involve Social Scientists in Defining the Anthropocene." *Nature* 540:7632 (2016): 192–93.

Fanon, Franz. *The Wretched of the Earth*, translated by Richard Philcox. Cape Town: Kwela Books, 2017.

Fowl, Stephen. *Idolatry.* Waco: Baylor University Press, 2019.

Francis (Pope). *Laudato Si': On Care for Our Common Home.* Vatican City: Encyclical Letter, 2015.

Galilea, Segundo. *Spirituality of Hope.* Maryknoll: Orbis, 1988.

Galtung, Johan. "Violence, Peace and Peace Research." *Journal of Peace Research* 6:3 (1969): 167–91.

———. *Peace by Peaceful Means: Peace and Conflict, Development and Civilization.* Oslo: International Peace Research Institute, 1996.

Gardiner, Stephen. *A Perfect Moral Storm: The Ethical Tragedy of Climate Change.* Oxford: Oxford University Press, 2011.

Ghosh, Amitav. *The Great Derangement: Climate Change and the Unthinkable.* Chicago: University of Chicago Press, 2016.

Gonzalez, Carmen G. "Bridging the North-South Divide: International Environmental Law in the Anthropocene." *Environmental Law Review* 32:2 (2015): 407–34.

Gorringe, Timothy. *The Education of Desire.* London: SCM Press, 2001.

Goudzwaard, Bob and Craig B. Bartholomew. *Beyond the Modern Age: An Archaeology of Contemporary Culture.* Downers Grove: IVP Academic, 2017.

———. "Economie als Afgod." In *Economie als Afgod? Kernvragen van het Conciliar Proces*, edited by Herbert.van Erkelends, 21–31. Kampen: JH Kok, 1989.

———. *Idols of Our Time.* Downers Grove: InterVarsity Books, 1989.

Granberg-Michaelson, Wesley. *Redeeming the Creation: The Rio Earth Summit: Challenges for the Churches.* Geneva: World Council of Churches, 1992.

Grey, Mary. *Sacred Longings: The Ecological Spirit and Global Culture.* Minneapolis: Fortress, 2004.

Grinspoon, David. *Earth in Human Hands: Shaping Our Planet's Future.* New York: Grand Central, 2016.

Guyer, Paul. "Introduction: The Starry Heavens and the Moral Law." In *The Cambridge Companion to Kant and Modern Philosophy*, 1–27. Cambridge: Cambridge University Press, 2006.

Hall, Douglas John. *The Cross in our Context: Jesus and the Suffering World.* Minneapolis: Fortress, 2003.

Hamilton, Clive. "Geoengineering: Our Last Hope, or a False Promise?" *New York Times*, May 26, 2013.

———. "Human Destiny in the Anthropocene." In *The Anthropocene and the Global Environmental Crisis*, edited by Clive Hamilton, Christophe Bonneuil and François Gemene, 32–43. London and New York: Routledge, 2015.

———. "The New Environmentalism will Lead Us to Disaster." *Scientific American*, June 19, 2014. https://www.scientificamerican.com/article/the-new-environmentalism-will-lead-us-to-disaster/ (accessed March 2, 2019).

———. "The Theodicy of the 'Good Anthropocene'." *Environmental Humanities* 7 (2015): 233–38.

———. *Defiant Earth: The Fate of Humans in the Anthropocene.* Cambridge: Polity Press, 2017.

———. *Earthmasters: The Dawn of the Age of Climate Engineering.* New Haven: Yale University Press, 2013.

———. *Requiem for a Species: Why We Resist the Truth about Climate Change.* London and New York: Routledge, 2015.

Hamilton, Clive and Jacques Grinevald. "Was the Anthropocene Anticipated?" *The Anthropocene Review* 2:1 (2015): 59–72.

Hamilton, Clive, Christophe Bonneuil and François Gemene (eds). *The Anthropocene and the Global Environmental Crisis.* London and New York: Routledge, 2015.

Hansen, James. "Game Over for the Climate." https://www.nytimes.com/2012/05/10/opinion/game-over-for-the-climate.html (accessed September 13, 2018).

Harari, Yuval. *Homo Deus: A Brief History of Tomorrow.* London: Vintage, 2017.

Haraway, Donna J. "Staying with the Trouble: Anthropocene, Capitalocene, Chthulucene." In *Anthropocene or Capitalocene? Nature, History and the Crisis of Capitalism*, edited by Jason W. Moore, 34–76. Oakland: PM Press, 2016.

———. *Staying with the Trouble: Making Kin in the Chthulucene.* Durham: Duke University Press, 2017.

Hartman, Laura M. "Healing the Climate? Christian Ethics and Medical Models for Climate Engineering." In *Theological and Ethical Responses to Climate Engineering,* edited by Forrest Clingerman and Kevin J. O'Brien, 129–48. Lanham: Lexington, 2016.

Hecht, Gabrielle. "The African Anthropocene." *Aeon Newsletter*, February 6, 2018. https://aeon.co/essays/if-we-talk-about-hurting-our-planet-who-exactly-is-the-we (accessed July 17, 2018).

Hefner, Philip. *The Human Factor: Evolution, Culture and Religion.* Minneapolis: Fortress Press, 1993.

Herbert, Brian. "A Conceptual Analysis of Abusive Charismatic Leadership." M.Phil. mini-thesis, University of the Western Cape, 2008.

Heuser, Andreas. "Charting African Prosperity Gospel Economies." *HTS Theological Studies* 72:4 (2016): 1–9.

Institute for Contextual Theology. *The Road to Damascus: Kairos and Conversion.* Johannesburg: Skotaville, 1989.

Jamieson, Dale. "Ethics, Public Policy, and Global Warming." *Science, Technology, and Human Values* 17:2 (1992): 139–53.

———. *Reason in a Dark Time: Why the Struggle against Climate Change Failed and What it Means for Our Future.* New York: Oxford University Press, 2017.

Jansson, Åsa. "Ecological Footprints." In *Global Change and the Earth System*, edited by Will Steffen et al., 126–27. Berlin: Springer, 2004.

Jenkins, Willis, Evan Berry, and Luke Beck Kreider. "Religion and Climate Change." *Annual Review of Environment and Resources* 43 (2018): 9.1–9.24.

———. *The Future of Ethics: Sustainability, Social Justice, and Religious Creativity.* Washington: Georgetown University Press, 2013.

Jones, Gregory. "A Thirst for God or Consumer Spirituality? Cultivating Disciplined Practices of Being Engaged by God." *Modern Theology* 13:1 (1997): 1–26.

Kinghorn, Johann (ed.). *Die NG Kerk en Apartheid.* Braamfontein: Macmillan, 1986.

Klein, Naomi. *This Changes Everything: Capitalism vs the Climate.* New York et al.: Simon & Schuster, 2014.

Kolbert, Elizabeth. *The Sixth Extinction: An Unnatural History.* New York: Picador, 2014.

Korten, David C. "Sustainability and the Global Economy." In *Visions of a New Earth: Religious Perspectives on Population, Consumption and Ecology*, edited by Harold Coward and Daniel C. Maguire, 29–42. New York: State University of New York Press, 2000.

Küng, Hans. *Eternal Life?* London: Collins, 1984.

Latour, Bruno. "Love Your Monsters: Why We Must Care for Our Technologies As We Do Our Children." *Breakthrough Journal* 2 (2011).

———. *Down to Earth: Politics in the New Climatic Regime.* Cambridge: Polity, 2018.

———. *Facing Gaia: Eight Lectures on the New Climate Regime.* Cambridge: Polity Press, 2017.

Lazarus, Richard James. "Super Wicked Problems and Climate Change: Restraining the Present to Liberate the Future." *Cornell Law Review* 94:5 (2009): 1153–233.

Leatt, James, Theo Kneifel and Klaus Nürnberger (eds). *Contending Ideologies in South Africa.* Cape Town: David Philip / Grand Rapids: Eerdmans, 1986.

Levin, Kelly et al. "Overcoming the Tragedy of Super Wicked Problems: Constraining our Future Selves to Ameliorate Global Climate Change." *Policy Sciences* 45:2 (2012): 123–52.

Lewis, Simon L. and Mark A. Maslin. *The Human Planet: How We Created the Anthropocene.* Penguin Books, 2018.

Long, D. Stephen. *Divine Economy: Theology and the Market.* London: Routledge, 2000.

Lovelock, James. *The Vanishing Face of Gaia: A Final Warning.* New York: Basic Books, 2009.

Lowes, Anthony. "Up Close and Personal: In the End, Matter Matters." In *Earth Revealing – Earth Healing: Ecology and Christian Theology*, edited by Denis Edwards, 125–44. Collegeville: The Liturgical Press, 2001.

Loy, David R. "The Religion of the Market." *Journal of the American Academy of Religion* 65:2 (1997): 275–90.

Lynas, Mark. "Geo-engineering, Nuclear Power and Climate Change: Playing God is Good for the Planet." *The Telegraph,* July 12, 2011.

———. *Six Degrees: Our Future on a Hotter Planet.* London et al.: Harper Perennial, 2008.

———. *The God Species: Saving the Planet in the Age of Humans.* Washington: National Geographic Society, 2011.

Maimela, Simon. "An Anthropological Heresy: A Critique of White Theology." In *Apartheid is a Heresy*, edited by John W. de Gruchy and Charles Villa-Vicencio, 48–58. Cape Town: David Philip, 1983.

Malm, Andreas and Alf Hornborg. "The Geology of Mankind? A Critique of the Anthropocene Narrative." *The Anthropocene Review* 1:1 (2014): 62–69.

Mamdani, Mahmood. "Reconciliation without Justice." *Southern African Review of Books* 46 (December 1996): 3–5.

Manenschijn, Gerrit. *Geplunderde Aarde, Getergde Hemel: Ontwerp voor een Christelijke Milieu-ethiek.* Baarn: Ten Have, 1988.

Maris, Virginie. "Back to the Holocene: A Conceptual, and Possibly Practical, Return to a Nature not Intended for Humans." In *The Anthropocene and the Global Environmental Crisis*, edited by Clive Hamilton, Christophe Bonneuil and François Gemene, 123–33. London and New York: Routledge, 2015.

Marquardt, Jens. "Worlds Apart? The Global South and the Anthropocene." In *The Anthropocene Debate and Political Science*, edited by Thomas Hickmann and Lena Partzsch, 200–18. London: Routledge, 2018.

Mauelshagen, Franz. "Bridging the Great Divide: The Anthropocene as a Challenge to the Social Sciences and Humanities." In *Religion in the Anthropocene*, edited by Celia E. Deane-Drummond, Sigurd Bergmann, and Markus Vogt, 87–101. Eugene: Cascade Books, 2017.

Mbembe. Achille. "Decolonizing Knowledge and the Question of the Archive." https://wiser.wits.ac.za/system/files/Achille%20Mbembe%20-%20Decolonizing%20Knowledge%20and%20the%20Question%20of%20the%20Archive.pdf (accessed November 30, 2018).

McDaniel, Jay B. *Living from the Center: Spirituality in an Age of Consumerism.* St Louis: Chalice Press, 2000.

McDonald, Evy. "Spending Money as if Life Really Mattered." In *Simpler Living, Compassionate Life: A Christian Perspective,* edited by Michael Schut, 50–66. Denver: The Morehouse Group, 1999.

McFague, Sallie. *Life Abundant: Rethinking Theology and Economy for a Planet in Peril.* Minneapolis: Fortress, 2001.

———. *The Body of God: An Ecological Theology.* Minneapolis: Fortress Press, 1993.

McKibben, Bill. *Falter: Has the Human Game Begun to Play Itself Out?* New York: Henry Holt and Company, 2019.

———. *The End of Nature.* New York et al.: Penguin, 1989.

McNeill, John R. and Peter Engelke. *The Great Acceleration: An Environmental History of the Anthropocene Since 1945.* Cambridge: The Belknap Press, 2014.

Meadows, Donella H. et al. *The Limits to Growth: A Report for the Club of Rome's Project on the Predicament of Mankind.* New York: Universe Books, 1972.

Meadows, Donella H., Dennis L. Meadows and Jorgen Randers. *Beyond the Limits: Confronting Global Collapse, Envisioning a Sustainable Future.* Post Mills: Chelsea Green, 1992.

Meeks, M. Douglas. "The Future of Theology in a Commodity Society." In *The Future of Theology: Essays in Honor of Jürgen Moltmann*, edited by Miroslav Volf, Carmen Krieg, and Thomas Kucharz, 253–66. Grand Rapids: Eerdmans, 1996.

———. *God the Economist: The Doctrine of God and Political Economy.* Minneapolis: Fortress, 1989.

Merchant, Carolyn. *The Death of Nature: Women, Ecology, and the Scientific Revolution.* San Francisco: HarperOne, 1980.

Messer, Neil. "Sin and Salvation." In *Systematic Theology and Climate Change: Ecumenical Perspectives*, edited by Michael Northcott and Peter M. Scott, 124–40. New York: Routledge, 2014.

Meyer, Brigitte. "Christianity in Africa: From African Independent to Pentecostal-Charismatic Churches." *Annual Review of Anthropology* 33 (2004): 447–74.

Midgley, Mary. *Beast and Man: The Roots of Human Nature.* Ithaca: Cornell University Press, 1978.

———. *Wickedness: A Philosophical Essay.* New York: Routledge, 2001.

Mische, Patricia. "The Integrity of Creation: Challenges and Opportunities for Praxis." In *Christianity and Ecology: Seeking the Well-being* of *Earth and Humans*, edited by Dieter T. Hessel and Rosemary Radford Ruether, 591–602. Cambridge: Harvard University Press, 2000.

Möller, Ulrich. "*Status Confessionis*? Confessing our Faith in the Context of Economic Justice." *Reformed World* 46 (1996): 138–44.

Moltmann, Jürgen. *God for a Secular Society: The Public Relevance of Theology.* Minneapolis: Fortress, 1999.

Moore, Jason W. "The Rise of Cheap Nature." In *Anthropocene or Capitalocene? Nature, History and the Crisis of Capitalism,* edited by Jason W. Moore, 78–115. Oakland: PM Press, 2016.

Morton, Timothy. *Dark Ecology: For a Logic of Future Coexistence.* New York: Columbia University Press, 2016.

———. *Hyperobjects: Philosophy and Ecology after the End of the World.* Minneapolis: University of Minnesota Press, 2013.

Nash, James A. "Towards the Ecological Reformation of Christianity." *Interpretation* 50:1 (1996): 5–15.

Neiman, Susan. *Evil in Modern Thought: An Alternative History of Philosophy.* Princeton and Oxford: Princeton University Press, 2002.

Nolan, Albert. "Theology in a Prophetic Mode." In *Hammering Swords into Ploughshares. Essays in Honour of Archbishop Mpilo Desmond Tutu*, edited by Buthi Tlhagale and Itumeleng Mosala, 131–40. Johannesburg: Blackshaws, 1986.

Northcott, Michael. "Eschatology in the Anthropocene." In *The Anthropocene and the Global Environmental Crisis,* edited by Clive Hamilton, Christophe Bonneuil, and François Gemene, 100–11. London and New York: Routledge, 2015.

———. "On Going Gently into the Anthropocene." In *Religion in the Anthropocene*, edited by Celia E. Deane-Drummond, Sigurd Bergmann and Markus Vogt, 19–34. Eugene: Cascade Books, 2017.

———. *A Moral Climate: The Ethics of Global Warming*. Maryknoll: Orbis Books, 2007.

O'Brien, Kevin J. "'First be Reconciled': The Priority of Repentance in the Climate Engineering Debate." In *Theological and Ethical Responses to Climate Engineering,* edited by Forrest Clingerman and Kevin J. O'Brien, 187–207. Lanham: Lexington, 2016.

———. *The Violence of Climate Change: Lessons of Resistance from Nonviolent Activists.* Washington, DC: Georgetown University Press, 2017.

Oreskes, Naomi and Erik Conway. *The Collapse of Western Civilization: A View from the Future.* New York: Columbia University Press, 2014.

———. *Merchants of Doubt.* London: Bloomsbury, 2010.

Palsson, Gissli et al. "Reconceptualizing the 'Anthropos' in the Anthropocene: Integrating the Social Sciences and Humanities in Global Environmental Change Research." *Environmental Science and Policy* 28 (2013): 3–13.

Pasztor, Janos. "The Need for Governance of Climate Geoengineering." *Ethics & International Affairs* 31:4 (2017): 419–30.

Peralta, Athena and Rogate Mshana (eds). *The Greed Line: Tool for a Just Economy.* Geneva: WCC, 2016.

Peters, Ted F. *Fear, Faith and the Future.* Minneapolis: Augsburg, 1980.

Pihkala, Panu. "Eco-Anxiety, Tragedy, and Hope: Psychological and Spiritual Dimensions of Climate Change." *Zygon* 53:2 (2018): 545–69.

Pinker, Steven. *Enlightenment Now: The Care for Reason, Science, Humanism and Progress.* London: Penguin, 2018.

Plantinga, Cornelius. *Not the Way it's Supposed to Be: A Breviary of Sin.* Grand Rapids: Eerdmans, 1995.

Potts, Rick. "What Will it Mean to be Human?" In *Living in the Anthropocene: Earth in the Age of Humans*, edited by W. John Kress and Jeffrey K. Stine, 27–31. Washington, DC: Smithsonian Books, 2017.

Raiser, Konrad. "Towards Defining the Concept of Structural Greed." In *The Greed Line: Tool for a Just Economy*, edited by Athena Peralta and Rogate Mshana, 91–93. Geneva: WCC, 2016.

Rasmussen, Larry. "The Future Isn't What It Used to Be: 'Limits to Growth' and Christian Ethics." *Lutheran Quarterly* 27 (1975): 101–11.

———. *Earth Community, Earth Ethics.* Maryknoll: Orbis, 1996.

Richardson, Neville. "Not Another Moral Summit! Problems and Possibilities for Moral Regeneration." *Scriptura* 82 (2003): 3–14.

Ricoeur, Paul. *Time and Narrative* (three volumes). Chicago: University of Chicago Press, 1984, 1985, 1988.

Rieger, Joerg. *Jesus vs. Caesar: For People Tired of Serving the Wrong God.* Nashville: Abingdon, 2018.

Ripple, William J. et al. "World Scientists' Warning of a Climate Emergency." *BioScience* biz088 (November 5, 2019).

Rittel, Horst and Melvin Webber. "Dilemmas in a General Theory of Planning." *Policy Sciences* 4 (1973): 155–69.

Rockström, Johan et al. "A Roadmap for Rapid Decarbonisation." *Science* 355:6331 (2017): 1269–71.

———. "Planetary Boundaries: Exploring the Safe Operating Space for Humanity." *Ecology and Society* 14:2 (2009): 1–32.

———. "Three Steps to Meeting the Climate and Nature Emergency." *The Telegraph*, July 16, 2019. https://www.telegraph.co.uk/business/how-to-be-green/climate-and-nature-emergency/ (accessed May 16, 2020).

Rolston, Holmes (III). *Conserving Natural Value.* New York: Columbia University Press, 1994.

———. *Environmental Ethics: Duties to and Values in the Natural World.* Philadelphia: Temple University Press, 1988.

———. *Three Big Bangs: Matter-Energy, Life, Mind.* New York: Columbia University Press, 2010.

Russell, Robert John, Nancey Murphy and William R. Stoeger (eds). *Scientific Perspectives on Divine Action: Twenty Years of Challenge and Progress.* Berkeley: Center for Theology and the Natural Sciences / Vatican City: Vatican Observatory Publications, 2008.

Sachs, Jeffrey. *Common Wealth: Economics for a Crowded Planet.* New York: Penguin, 2008.

———. *The End of Poverty: Economic Possibilities for Our Times.* New York: Penguin, 2005.

Sagoff, Mark. "A Theology for Ecomodernism: What Is the Nature We Seek to Save?" *Breakthrough Journal* 5 (2015).

Schumacher, Ernst F. *A Guide for the Perplexed.* New York: Harper & Row, 1977.

Scott, Peter M. "God's Work Through the Emergence of Humanity." In *T&T Clark Handbook on Christian Theology and Climate Change*, edited by Ernst M. Conradie and Hilda Koster, 373–83. London: T&T Clark, 2019.

Scranton, Roy. *Learning to Die in the Anthropocene: Reflections on the End of a Civilization.* San Francisco: City Light Books, 2015.

———. *We're Doomed. Now What? Essays on War and Climate Change.* New York: Soho Press, 2018.

Serres, Michel. *Hominescence.* London: Bloomsbury, 2019.

———. *Malfeasance: Appropriation through Pollution.* Stanford: Stanford University Press, 2011.

———. *The Incandescent.* London: Bloomsbury, 2018.

———. *The Natural Contract.* Ann Arbor: The University of Michigan Press, 1995.

Skrimshire, Stefan. "Rewriting Morality: A Theological Critique of Geoengineering and De-Extinction." In *Theological and Ethical Responses to Climate Engineering*, edited by Forrest Clingerman and Kevin J. O'Brien, 103–26. Lanham: Lexington, 2016.

Sloterdijk, Peter. *In the World Interior of Capital: For a Philosophical Theory of Globalization.* Cambridge: Polity Press, 2014.

Smit, Dirk J. "The Symbol of Reconciliation and Ideological Conflict in South Africa." In *Reconciliation and Reconstruction*, edited by Willem S. Vorster, 79–112. Pretoria: Unisa, 1986.

———. "What Does *Status Confessionis* Mean?" In *A Moment of Truth: The Confession of the Dutch Reformed Mission Church*, edited by G. Daniel Cloete and Dirk J. Smit, 7–32. Grand Rapids: Eerdmans, 1984.

Smith, Nico J. "Apartheid in South Africa as a Sin and Heresy: Some of its Roots and Fruits." In *New Faces of Africa: Essays in Honour of Ben (Barend Jacobus) Marais*, edited by Jan W. Hofmeyr and Willem S. Vorster, 143–52. Pretoria: Unisa, 1984.

South African Council of Churches. *Climate Change – A Challenge to the Churches in South Africa*. Marshalltown: SACC, 2009.

Starke, Linda (ed.). *State of the World 2004.* Washington: Worldwatch Institute, 2004.

Steffen, Will et al. "Planetary Boundaries: Guiding Human Development on a Changing Planet." *Science* 347:6223 (2015): 736–47.

———. "Trajectories of the Earth System in the Anthropocene." *Proceedings of the National Academy of Sciences of the United States of America* (August 6, 2018): 1–8.

———. *Global Change and the Earth System: A Planet under Pressure.* Berlin: Springer, 2004.

Steffen, Will, Jacques Grinevald, Paul Crutzen, and John McNeill. "The Anthropocene: Conceptual and Historical Perspectives." *Philosophical Transactions of the Royal Society A* 369:1938 (2011): 842–67.

Stringfellow, William. "Money." In *Simpler Living Compassionate Life: A Christian Perspective*, edited by Michael Schut, 67–72. Denver: Morehouse Group, 1999.

Sung, Jung Mo. "Greed, Desire and Theology." In *The Greed Line: Tool for a Just Economy*, edited by Athena Peralta and Rogate Mshana, 43–54. Geneva: WCC, 2016.

———. "Toward a Conceptual Elaboration of Structural Greed." In *The Greed Line: Tool for a Just Economy*, edited by Athena Peralta and Rogate Mshana, 94–99. Geneva: WCC, 2016.

———. *Desire, Market and Religion.* London: SCM Press, 2007.

Sutcliffe, Bob. "Development after Ecology." In *The North, the South and the Environment: Ecological Constraints and the Global Economy,* edited by Vinit Bhaskar and Andrew Glyn, 32–58. London: Earthscan Publications, 1995.

Swilling, Mark and Eve Annecke. *Just Transitions: Explorations of Sustainability in an Unfair World.* Cape Town: UCT Press, 2012.

Symons, Jonathan. "Geoengineering Justice: Who Gets to Decide Whether to Hack the Climate." *Breakthrough Journal* 9 (2018).

Taylor, Michael. "On Greed: Toward 'Concrete and Contemporary Guidance for Christians'." In *The Greed Line: Tool for a Just Economy*, edited by Athena Peralta and Rogate Mshana, 80–98. Geneva: WCC, 2016.

Ten Bos, René. *Dwalen in het Antropoceen.* Amsterdam: Boom, 2017.

Terreblanche, Sampie. *A History of Inequality in South Africa 1652–2002.* Scottsville: UKZN Press, 2002.

Toulmin, Stephen. *Cosmopolis: The Hidden Agenda of Modernity.* New York: The Free Press, 1990.

Tucker, Mary Evelyn and John Grim. "Series Foreword." In *Christianity and Ecology: Seeking the Well-being* of *Earth and Humans*, edited by Dieter T. Hessel and Rosemary Radford Ruether, xv–xxxii. Cambridge: Harvard University Press, 2000.

Tutu, Desmond. "Christianity and Apartheid." In *Apartheid is a Heresy*, edited by John W. de Gruchy and Charles Villa-Vicencio, 39–47. Cape Town: David Philip, 1983.

Van den Brink, Gijsbert. "Should we Drop the Fall? On Taking Evil Seriously." In *Strangers and Pilgrims*, edited by Eddy Van der Borght and Paul van Geest, 761–77. Leiden: Brill, 2011.

Van der Ven, Johannes. *Formation of the Moral Self.* Grand Rapids: Eerdmans, 1998.

Van Hoogstraaten, Hans Dirk. *Deep Economy: Caring for Ecology, Humanity and Religion.* Cambridge: James Clarke & Co., 2001.

———. *Geld en Geest: Over Milieu-ethiek.* Baarn: Ten Have, 1993.

Van Huyssteen, J. Wentzel. "Questions, Challenges, and Concerns for the Image of God." In *Finding Ourselves after Darwin*, edited by Stanley P. Rosenberg, 33–47. Grand Rapids: Baker Academic, 2018.

———. *Alone in the World: Human Uniqueness in Science and Theology.* Grand Rapids: Eerdmans, 2006.

Wagner, Frank. *Geld oder Gott? Zur Geldbestimmtheit der kulturellen und religiösen Lebenswelt.* Stuttgart: Ernst Klett Verlag, 1984.

Wallace-Wells, David. *The Uninhabitable Earth: Life after Warming.* New York: The Duggan Books, 2019.

Ward, Graham. *Cities of God.* New York: Routledge, 2000.

White, Carol Wayne. "The Wicked Problem of Climate Change: Re-Envisioning Hope: Anthropogenic Climate Change, Learned Ignorance, and Religious Naturalism." *Zygon* 53:2 (2018): 570–84.

White, Lynn. "The Historical Roots of our Ecological Crisis." *Science* 155 (1967): 1203–7.

Williston, Byron. *The Anthropocene Project: Virtue in an Age of Climate Change.* Oxford: Oxford University Press, 2015.

Wood, David. *Deep Time, Dark Times: On Being Geologically Human.* New York: Fordham University Press, 2019.

World Alliance of Reformed Churches. *Processus Confessionis: Process of Recognition, Education, Confession and Action Regarding Economic Injustice and Ecological Destruction.* Background Papers No. 1, 1997.

World Council of Churches. *Alternative Globalization Addressing Peoples and Earth.* Geneva: WCC, 2005.

———. *Solidarity with Victims of Climate Change.* Geneva: WCC, 2002.

Yusoff, Kathryn. *A Billion Black Anthropocenes or None.* Minneapolis: University of Minnesota Press, 2018.

Zalasiewicz, Jan et al. "When did the Anthropocene Begin? A Mid-twentieth Century Boundary Level is Stratigraphically Optimal." *Quaternary International* 383 (2015): 196–203.

Žižek, Slavoj. "Censorship Today: Violence, or Ecology, as New Opium for the Masses." https://www.lacan.com/zizecology1.htm (accessed December 10, 2019).

———. *Living in the End Times.* London and New York: Verso, 2011.

———. *Trouble in Paradise: From the End of History to the End of Capitalism.* Brooklyn: Melville, 2014.

Index

About the Author

Ernst M. Conradie is senior professor in the Department of Religion and Theology where he teaches systematic theology and ethics. His most recent monographs include *The Earth in God's Economy: Creation, Salvation and Consummation in Ecological perspective* (LIT Verlag, 2015) and *Redeeming Sin? Social Diagnostics amid Ecological Destruction* (Lexington Books, 2017). He is the coeditor (with Hilda Koster) of *The T&T Clark Handbook on Christian Theology and Climate Change* (T&T Clark, 2019).